Richard Cross

The Cotton Yarn Spinner Showing How the Preparation Should Be Arranged for Different Counts of Yarns

Vol. 1

Richard Cross

The Cotton Yarn Spinner Showing How the Preparation Should Be Arranged for Different Counts of Yarns
Vol. 1

ISBN/EAN: 9783337811679

Printed in Europe, USA, Canada, Australia, Japan

Cover: Foto ©Lupo / pixelio.de

More available books at **www.hansebooks.com**

THE

COTTON ⸸ YARN ⸸ SPINNER,

SHOWING

How the Preparation should be Arranged

—FOR—

DIFFERENT COUNTS OF YARNS,

—BY A—

System more uniform than has hitherto been practiced; by having a Standard Schedule from which we make all our Changes.

—BY—

RICHARD CROSS.

———

PUBLISHED BY THE AUTHOR,
No. 1717 NORTH FRONT STREET, PHILADELPHIA.
—1882—

CONTENTS.

INDEX TO TABLES.

PREFACE.

IN introducing this work to the reader, it is earnestly requested of him to have patience to read the introductory remarks, before he attempts to peruse its general contents, so that he may obtain a previous knowledge of why, and how, the writer desires he should become acquainted with the object of bringing before the Public such a book.

By a long experience with the machines, and the variety of yarns that can be made on them. I wish to show in a very concise manner, how they are made to perform such actions, by a thorough practical system of arranging and adjustments, which will produce in the first trial precisely what you want to do, and make, under the disposition of a competent man, for this work is not intended for Theorists, but men, who are perfectly acquainted with the machines, and Tyro's who are training under the discipline of their overseers, for it is dangerous and impolitic to the work and machinery to tolerate the unacquainted meddling with important and expensive machines, which should be handled by a mechanic, and not by empirics, who have the assumption to dictate and know more in a few months, than those who have labored a whole lifetime to secure a higher position, by the knowledge they have acquired so assiduously.

I have often wished for some one to introduce a more improved system of making different counts of yarn, by which the manufacturer can make his changes, from a schedule, having become impatient by the want of such a formula I undertook the task myself, in which you will find extracts from the general schedule put in a table form to suit the machine, and a place for it in the book.

When looking over these tables for the purpose of changing your counts, you will soon discover how consistent, is the preparation to the Numbers of Yarn.

Now this being the object, which induced me to write this work, that every manufacturer, individually and collectively, might, if he chooses, adopt this method, if he finds it the nearest and most economic manner, out of the many different ways of producing the numbers required, for there is only one way that is perfect under every advantage, and this we are trying to emulate, and in so doing, it is my earnest wish and desire that you will give this method a fair trial, that you may realize and be honestly convinced of its merits. If I am favored with such results that you feel confident of its success, there will be one more who has conceived the intelligent manner, which this system exhibits, and I feel confident will become general, that being so will be the climax of my ambition ' for it is natural I should feel so, having spent a lifetime in this business, acquiring knowledge by the experience and opportunities afforded me, am willing now, for the benefit of those who are inclined to be progressive, to turn over this accumulated capital for the general welfare of machinists and spinners, hoping they will invest it with as much care as I have used in acquiring it, trusting that the interest which will be derived from it may assume the proportion of my wishes.

With a view of reducing the price of this book, I have omitted the drawings and illustrations of various machines represented, they not being necessary for those whom this book is intended; also, the lengthy calculations which are superfluous in a book of this kind, but have given the RULES, so that any question may be solved on the slate in a very brief and reliable manner, showing in every instance how the work has been epitomized, where it could be admitted for the benefit of the purchaser, and yet it contains as much useful information to those who have the responsibility of making the yarns, and more practical suggestions that have not been given in other works on the same subject; submitting the same to your own judgment, by a strict perusal of its contents.

RICHARD CROSS.

MIXING.

IN commencing to write on the subject of making Yarns, our starting point will be from the bales of cotton we have received, these should be examined after taking the hoops and bandage off, spreading them open and selecting from them such as has been already classified according to the kinds of yarn you intend to make, for in making warp your stock will require to be of a more even staple, viz.: in length, and if you should have some bales that are not so good, but of about the same length, you may in making your mixing, put a certain percentage of this class, but not so much as to reduce the strength of your yarn too much. You will first take from No. 1 bale and spread out on the floor, occupying about as much space as you need for the quantity you intend to mix. Then take from No. 2 bale of another quality and spread on the top of the first layer, and so on alternately, but if you have selected three kinds for your mixing, then you will take and spread them from Nos. 1, and 2, and 3, then from Nos. 1, and 2, and 3, and so on, keeping in view the space you have at command. This mixing must be regulated by the capacity of your mill and it would be well if you did not exceed over 5 feet in height, so that in taking it from the mixing to the opener you will be able to handle it much better, as it should be taken from top to bottom so that the classes may be kept proportionately mixed. Now this mixing is supposed to be for making warps, for which I proposed an average length of staple, and not to exceed in length the kind you intend to use generally and regularly.

Having once made your mind up as to the numbers you intend to spin, your class of cotton can then be determined on, to secure uniformity of thread and the adjusting of the preparatory machines. You will find out by this mode of mixing that it will give the cotton a better chance of spreading and opening out by its elasticity and absorbing atmospheric moisture, and if need be, sometimes a little artificial will do no harm to strengthen if not used too soon after employing it. But these are measures not easily dealt with if not cautiously attempted, but act disastrously if not done expertly and by a trustworthy person, on whom you can rely for doing it properly, according to your instructions, but really should be made under your supervision if time will admit. I would also prefer cotton that has been baled for a length of time, for *new cotton* does not work so well as old, everything being equal, new cotton being too lively and elastic, or lofty, which makes it difficult to prepare, on account of its non-condensing quality which is a great art in preparing cotton.

This artificial mode of moisture I have already spoken of, is, by using water, put on each layer of cotton when mixing by a sprinkling can having a very fine rose, and, if by a strict measure to a certain area, you will be able to ascertain just the right quantity after a few exact trials, being governed by its facility of working and quality of production, you can determine by its results what kind of a specific it really is, to adopt when necessary.

There are other modes of mixing cotton besides this, but they are not so good and useful in their effects: for instance, if you take a bale of cotton and run it through the opener and make it into laps on the spreader, then take another bale and do the same, and so on. Then you put one of each kind on Lapp machine, which is a very ready method of mixing ; but you lose the most important part, *before mentioned,* and which I wish to impress on your minds as very essential and an active principal in mixing cotton. It may not appear to those who have adopted the latter mode of mixing, of having the economy which they claim, by dispensing with mixing on the floor, but I maintain that the first plan I explained is the best, both for the machines, and the better for it being ready of itself for manipulating. Here you see we gain two points over the other plan and will be better cleaned, for it is here in this room that we expect to leave all the trash and dirt, for what is done here should be well done, and not leave it to be carried on to another machine or card, before all the hard and tough substances have been driven out by the beaters, so as not to injure the card wire when it has progressed so far. This is a most important point and is the compound of the other two points, giving us a cleaner and firmer lap, saving waste and labor in not having to carry back the waste made from broken and sticky laps being made too hurriedly, allowing no time for the fibres to expand after being released from the great pressure which was on it in the bale, but pitched right on to the machine in *lumps,* to be ripped and torn from the feed rollers, having more pressure on them than where it leaves them almost without being beat at all, by being so unevenly spread on the apron and passing under the feed rollers *thick* and *thin,* or *solid and puffy,* which is complete murder. Hence the rough and reckless laps, which could have been avoided by a proper mixing being done according to an approved system of long standing, and evidently the best plan when exhibited by experiment, showing clearly a claim which I prefer.

On the other hand, by No. 2 plan, it is claimed they are

less liable to damage by fire, hence the premium on insurance will be less. Now I have some doubts respecting an allowance being made on that account, and whether the theory it presents before the Insurance Agent, is not a delusion, taking a *prima-facia* view of it, as they generally do, and whether it will stand as secure against fire as No. 1 plan *will*, after a proper investigation. It is very seldom a fire takes place, only by some foreign matter of an iron or flinty nature, being struck by the beaters, which must certainly have escaped detection by the handling of the cotton before passing under the feed rollers. Now such an accident might happen by either plan, but if this be the chief cause of fires, I am fully convinced that No. 1 plan of mixing will have the preference, from the very fact of its having a better chance of being detected by its greater handling, shaking and tossing in laying the mixing on the floor, and then being taken up again to spread on the feed apron. I will submit to the contrary only—if any Insurance Agent that has had great practice and opportunities of testing and averaging the numbers of fires and how they are ignited, whether from No. 1 or from No. 2 plan has had the least accidents by fire.

Although it is safe to say having once got on fire, the damage is not so risky when the cotton is in the bale, as it is when spread out in a mixing ; yet, with the modern appliances for extinguishing these fires, it would be a rare occurrence indeed, if it should make any headway during working hours, and it must happen then, taking our argument through from a mixing point. But I am wandering away from my subject, for it is not the insurance I am treating on, but to look for a better class of yarn than is sent to market, which can only be made by clearing away some of the novelties advanced in the present decade, which have deteriorated more than excelled in the manufacturing of them, and more especially, at this part of it which we are now considering, viz: mixing and opening and the variety of devices employed on these machines to do it, but they are gradually wearing out, as well as the new ideas they promulgated in their favor, are vanishing, when meeting a more formidable and established practice, and entertained through long experience by being tested side by side with the facts on one side and the ideas on the other, have placed the former in a more permanent use by those who have experimented and chosen a more radical, than visionary view as to the choice of machines.

I will again remark in choice of stock for making good warps, it must be an average length of staple when they are made by a continuous spinning—such as Ring & Flyer, and cap Frames, and the uniformity of the thread is in accord with the uniformity of staple, of cotton, so it would not be very wise to use *fly* or *short* weak cotton in the mixing, unless they are very coarse yarns, and then very sparingly.

Now, in mixing filling, or weft, or hosiery yarns the mixing may be more general in the use of different grades from the same class of cotton, according to the numbers of yarn and its adaptability, for the filling for sheetings we could not pretend to use the same class of cotton as would be required for making hosiery yarn, so it will be a matter of judgment how far you can encroach with the lower grades and keep your goods up to the standard, or having the same features one time as at any other time, by which if you have acquired a good sample, then hold on to that system, and you will surely succeed if you have made a moderate margin be-

tween the raw cotton and the goods. In your mixing for
this kind of yarn you have an advantage over warp yarn
in using inferior stock, and in the use of intermittent spinning, called the mule, which assists by a propensity belong-
ing to it, to recover by this act of stretching the unevenness
of the thread caused by a mixture of unequal fibres. In
mixing for hosiery yarns, you should choose a clean and
spiral natured class, so as to give it elasticity, and make it
bulky; the South American cotton is the best for this kind
of yarns, such as Perus, Marahnams, Bahia and (New
Orleans and good Mobile). I think the two latter will
suit for the kind of hosiery yarns for this market, as the
fineness is not absolute here so much as cheapness with the
laboring class of people.

The same class of cotton I would use for making *fine
warps*, and for a coarser kind I would mix Tennessee and
Middling, Upland and Texas, for 16's up to No. 20's warps.
But in mixing for filling or weft, use Low Middling, Up-
lands and Texas, and grade them according to numbers of
yarn you are going to make, allowing a percentage of *Fly*,
if you choose to use some of your own from double carding only. Also a per cent. of your waste which you will
have to use up by putting in the mixing only, as it would
be injurious to use it indiscriminately in the making of
good yarns, or even salable yarns. This method of mixing promiscuous lots of cotton together, which I have been
in the habit of seeing at various mills, is a complete
slaughter of what would probably have been a salable
article, if the mixing had been made with some kind of
ordinary judgment, instead of rushing through anyhow,
but which is usually done by the hands working in the
department, being indifferent to the responsibility placed
upon them for making good laps instead of seeing that

they are neat and well made, having smooth selvedges and properly condensed, so that no waste from that source returns to be battered up again to the loss of the manufacturer, by weakening the staple and waste of time and labor by this repetition.

OPENING.

Now we will begin to open this cotton from the mixing already made, by taking it from top to bottom in your arms, and then spreading it on the apron of the machine, or in the funnel, according to the construction of the machines, of which there is a great variety, and this I will leave to your choice from your own good judgment to select, making preference to them that consume the least power, and leaves the cotton in a very loose and open condition, so that it will spread even under the feed rollers ; this can be done best by a cylinder that has teeth laid transversely, making it a revolving rake. This kind of a beater is not so rigid, and will tease the lumps out more easily at its first initiation, and loosen it out by passing over the *grate bars* which are set to suit the grade of cotton you are using, which can be closed or opened wider, by an attachment outside of the machine, and showing by an indicator their inclination. You will judge for yourself by examining the droppings if there is any loss of cotton by coming through them; if there be too much cotton among them, the grate bars must be closed a little to prevent this waste, for in some mixing, there will be a great difference and should be inspected every day, and kept clear and clean, that all extraneous matter may not be retarded but allowed to be forced through by the force of the revolving rakes and their own gravity, and drop underneath and be carried away as often as it accumulates there, to prevent it from stopping up these useful appliances. The cotton is then allowed to pass on to the cylinder cages, either on the machine or else up through a wooden flue, extending across the room at the end, drawn by a *fan* by which during its passage, it tilts and dances over a false bottom made of lattice-work, under which are doors that can be unbuttoned, and taking out all the *leaf*, moats, and non-fibrous matter that has fallen from the loosened cotton on its way to the cage at the other end. This arrangement excels all previous devices, and is done automatically I may say, although not without power, as the *fan* requires considerable, but the cotton leaves its residue without any beating whatever, in leaving the cages here it falls down an upright trough or box which is made to conform with the feed-rollers, in such a manner as to become a self-feeding apparatus, and goes along under the feed roller, where it is subjected to a little more raking by the *first* beater, when it is now about prepared for the *rapid* beaters and made into a *lap*. These laps are then put on the finisher lap machine, the apron sides are made to receive three or four laps or doublings; here I would use only one beater, with *two* blades running 1300 revolutions, and the fan · 5500 feet per minute, and the speed of the cone or regulator or evener, should run about as many revolutions as` will give the feed roller a surface velocity of seven feet per minute, so with a draught of three on this machine as a finisher, the surface velocity of *lap* roller will be eighteen feet per minute or six yards—equal to a lap of 26 lbs. at 10-8 ozs. per yard in 6½ minutes, averaging 2000 lbs. of laps in 10 hours. This gives us some idea of the required weight, to be received from the opener or scutcher to keep the finisher supplied. Now the opener will require a less number of revolutions on the first porcupine, (as they are often called), on account of there being more blades or rakes to it, and then again, the feed apron will require to be slowered at a rate of four feet per minute, and the feed roller five feet per minute, with a weight of 37.⁵ ozs. spread on three feet of length of apron, unless fed automatically and then the weight will be the same.

I have reason to believe that the 1st beater or porcupine or cylinder will answer best with our cotton, with 6 of the rakes laid transversely on this cylinder running 800 revolutions per minute, this would give the fibres 4800 blows, equal to 100 licks to every inch delivered to it, and I should say that that is raking it pretty well for a start. As it advances to the next feed roller there will be another draught increasing in surface velocity, but the beaters here are two bladed with a constant speed of 1300 revolutions, reducing this beating as it progresses, to the lap, in ratio with the draught, and the speed of the lap roller is constant with the feed apron on this machine; it having no regulator attached to it. I may as well state here, that those who have charge of these machines should look to the grate bars being kept up to their places, they are usually set about 1 inch from beater blades, and ½ inch apart although these distances are not to be considered permanent, but are meant as an approximate to what is required. You will also see that the *fan* is properly placed underneath the cages, if it be an eccentric fan, then the longest radii of cover will be toward the feed roller and drive the air out toward the lap using an open belt for the fan and a strict attention to the draught regulators, so as to keep an evenly spread sheet before passing under the cages, seeing that all parts of the machine are made air tight, only what is necessary through the regulators for exhaustion by the fan, and this air should all pass through the grate bars and cages, the heavy substances being driven through the grate bars against the current of air going through them where the beater is dashing the cotton against the sharp edges of these bars, and making it into flakes and spongy fleeces, when passing along with the draught, it drops some of the lighter material, such as leaf, moat etc., through the bars which are of alternate depths and kind of arched at the tops, this convex side being toward the beater it has just left, and underneath these bars must be a box fitted so as to exclude any air from passing through them; the dirt which accumulates should be let out by a door fitted underneath, held up by two iron arched levers and retain their position by a weight being fastened on the same rod at right angles and by raising this up on the outside of machine, the arched lever falls down and the door resting on them, letting all the rubbish fall on the floor, which should be repeated several times a day. Having passed through the cages it now comes to be calendered, which is a most essential appliance for keeping good the work that has already been done, and securing to the laps a firmness of texture so they will unroll without making waste, and this can be done best by the four rollers over each other, and lever weighted; for spring and rubber attachments are a complete nuisance, for there is no regularity of pressure that can be obtained by them, and no solid substance can go under them without bursting them all up; and besides, we require precision in the amount of pressure at each end of the calenders, which can be acquired exactly by leverage, and rendered free from accidents. I prefer *sureties* to notions and guesses in such parts of a machine where there is such power required, by giving strength and durability to its component parts, also simplicity of construction, with a view to its economy in the value of these machines.

Now, we will suppose the machine to be placed in position, by being set convenient and in line with the *driving shaft*, all the other drivings are supposed to be fixed and permanent on the machine by the maker, with regard to their speeds and draughts, so as to need no mechanical contrivances applied after leaving the machine maker, for the manufacturer has not the facilities at command, neither does he expect to spend any more money on them than for the belting to drive them with. After setting the beaters, the belts can be put on and the journals limbered up, so that they will run smooth and steady without heating; for, by pouring oil and tallow on to prevent heating, the grease gets inside of the bearing and runs on the grates and beaters, and if not cleaned off before putting cotton through, the laps will be spoiled, by having no regularity of surface, which has been caused by the beater blades or axle having lumps of cotton hanging to them on this grease. When you have made this all right, you may then adjust the levers on the *fluted* rollers, leaving them ample room for rising and falling in case any undue variations to thickness may occur; and when you have them right, see to making them secure, to prevent unequal pressure, then the calenders may be weighted, leaving a good margin for altering, as they cannot be made secure until you have made a few laps, and seen how they unroll themselves at the card. When being satisfied that they are condensed enough only, for excess will require more power to drive them, you will then make the weight secure, so as not to slip on the lever. You see by this method of weighting there is no increase of weight on the calender if any hard substance should go under them, but simply rises and lets it go ; but where springs and rubbers are used, there is no relief, but extraordinary pressure, and increases in proportion to substance.

Then you will next examine the racks, and lap rollers, and friction pulleys that are enclosed on top of racks and see to their being perfectly free and easy, both racks working up and down freely and dropping simultaneously on their bearings, so that the pressure on the ends of lap roller will be equal, and producing a *lap* in shape of a true, solid cylindroid in appearance; there must also be strict attention paid to the *brake* before starting the machine, taking the precaution to set the weight on the end of the friction lever, so that you can turn the shaft by hand, when the *brake* is applied at first, for sometimes it gets neglected, and then by extra friction being on, the lap roller gets sprung and the wheels break, causing a great expense to remedy, especially if your mill is a long distance from the machine maker; but this can all be avoided by using the present advice. I may here state that it is much better to have a lap dish to roll your lap in when taking from it the roller, if the machine is so placed as to leave room for drawing it out, because by taking a lap and dumping it on a block it very often loosens the roller ends, which are costly to fix again, besides the lap is handled much better when the roller is drawn out in the dish preventing rough and bad selvedges. You will next direct your attention to the feed apron, having previously set your feed roller to Beater blades about 3-16 inches off; you can then adjust your apron sides square with the machine, so that the apron will run free and keep the slats or lattices from catching at the ends and ripping off. There is an arrangement attached to the sides, by which, with the extension screws and bracket, will, by moving in or out the end of the apron roller, bring it square and clear of catching; it is also useful for giving the apron the right tension, as it will get a little slack at

times by the lattice belts stretching a little. We will now
fasten the regulator to the machine, making the driving
cone axle plumb, when in gear with the feed roller. In
setting the *evener* for the weight of lap to be used there is
a *stud* behind the cone box which can be moved ; or the
regulating *lever* that moves the belt quadrant, has a long
slot in it which can be lengthened or shortened to suit the
variable weight, and by making the leverage shorter from
the stud to the belt shifter, it is not so liable to run the
belt off the cone, when the weight varies on the feed,
although it does not work so sensitive, as when the leverage is shorter on the other side over the screw.

REGULATOR OR EVENER.

The principle of the regulator is, that the apron is made
to vary its speed according to the thickness of cotton
spread on apron, that is intended to go between the feed
rollers ; for if it goes through thick, the apron goes slower.
The speed of *driving cone* is in proportion to the length.
moved by the belt shifter, these belt forks moving always
to keep the belt at right angles with the *cone shaft* and the
diameter of the two cones at the point where the belt is
running, when added together will always be a constant
sum.   Now, leaving a certain weight of cotton on *one yard*
of apron, will give a length equal to draught when the
cone belt is in the centre of cones, and to get the thickness will be to suppose the lap to weigh so much to *one*
(1) yard and to lose five per cent., then the weight and per
cent. of loss will have to be spread on the apron ; in
consequence, the top feed roller will be lifted up by
the thickness of cotton, and that should be equal to
centre of cone, or set so by long nut or by sett screws in

bracket at the end of lever; the weight lap can also be altered if you should require it, with the same draught by the resetting of cone belt in the centre, when the thickness of feed is under the rollers, the speed of apron is in inverse ratio to the thickness of cotton going through the rollers, or this multiplied by the revolution of the roller, geared in the driving cone will give the speed of apron. But generally in reviewing this evener for regulating, it simply amounts to this, that the lap roller velocity is constant, and the driving cone is constant, so when the cone belt is in the centre the draught is constant because the driving cone is constant.   Now suppose the lap should be a little too heavy, then the attendant goes and unscrews the long nut a little and that moves the cone belt from the centre, which causes the driven cone to be slowered, hence, an increase in the draught and the weight of lap decreased; but by doing so he has left a greater margin on one end of cone, and shorter on the other end, which should not be, when you have got the desired weight, so he will just reduce the weight spread on the apron and screw the long nut up again until the cone belt runs in centre again leaving room for variations in feeding, caused by laps running out, and sometimes going through double and single by which will be seen the utility of conjoining such a useful and almost indispensible piece of mechanism to the machine.   In examining two of these machines, made by different firms I found the draught of Taylor and Langs to be $\frac{19}{12}$ $\frac{40}{24}$ $\frac{26}{60}$ $\frac{48}{18}$ = 3.04 draught and Whitings to be $\frac{100}{44}$ $\frac{72}{33}$ $\frac{35}{32}$ $\frac{13}{104}$ $\frac{48}{13}$ = 2.53 draught.

We have now got the draught of the machines so we will see how to get the right weight of lap, I will refer to my previous statement which was 37.2 ozs. spread on 3 feet of apron then 3.04)37.2(equal 12.23 minus 5 per cent. equal 11.-61 ozs. the lap on first machine, we take 3 of these laps and put them on the feed apron of finisher lap machine, this is termed doubling, so 3 x 11.61 divided by 3.04 equal 11.45 minus 5 per cent. equal 10.8 oz. lap to *one* yard in length, this being according to the schedule for a certain number of counts made from a 36 inch lap in width having ascertained now all the chief requisites for making good laps we can now start up and I will venture to say go ahead; and we will now leave these machines and go to its successsor which is called a *card*.

No. 1 TABLE

——FOR A——

36-Inch Top Flat Card.—The Cylinder 1250 feet per Minute.—Doffer 420 Inches per Minute.

SINGLE CARDING.

DRAUGHT OF CARD 90.

Nos.	Hank Sliver.	Grains Per Yard.	Weight of Lap.	Weight of Lap.
4	.149	56	12.	Draught of Card 112
6	.154	54	11.8	
7	.157	53	11.7	
8	.160	52	11.57	
10	.164	50	11.36	
11	.168	49	10.25	
12	.173	48	10.15	12.9
13	.177	47	10.05	12.63
14	.181	46	9.87	12.36
15	.185	45	9.7	12.1
16	.189	44	9.45	11.83
17	.194	43	9.25	11.56
18	.198	42	9.	11.29
19	.203	41	8.8	11.02
20	.208	40	8.6	10.8

DRAUGHT OF CARD 112.

Nos.	Hank Sliver.	Grains Per Yard.	Weight of Lap.
21	.213	39	10.2
22	.218	38	10.07
23	.224	37	9.9
24	.231	36	9.65
25	.237	35	9.36
26	.245	34	9.1
27	.252	33	8.8
28	.260	32	8.55
29	.268	31	8.25
30	.276	30	8.

TOP FLAT CARD.

The card is a machine that follows the Lap finisher, when the *lap* is supposed to be made in a uniform thickness and width, so as to get as even a sliver as possible.

Before being presented to the carding process it should be understood what quality of cotton and how well it is prepared for the numbers of yarns you are going to spin, so that we may form some idea of the capacity of such a card to those counts ; it is economy to have such a foresight, so that we may know precisely how *much* work can be done (and not how little), also how well in accord with the *quantity*, for it is nearly time that some better system was, or should, be adopted, and this bragging stopped, of how much you are doing per day in excess of your neighbor ; it would be much better for all of us, by using some system approximating to a *law*, whereby we have through our practice and experience, proved that such (causes give such an effect) which substantially becomes a *law*, and must be referred to, so that we may more readily accomplish our work, which such a systematic course will enhance the value of it by its regularity, and its production by uniformity in price, with the numbers of counts made. I have considered this very often, and seriously ; and intend to make a schedule for the benefit of the manufacturer and his overseers, for reference, for it would be ruinous to the manufacturer to use the same material and carding for No. 10's that he is making for No. 20's, unless he is paid more for the extra expense of preparation, and then it looks to be troublesome to change mixing and weight of lap, etc.; " but this must be done " and adhere to the schedule at every point where there is shown to be a change of material and in the machines, making your business more profitable and the goods more salable, by having a system which will show for itself, by being methodical and in compliance to the schedule referred to; how much easier it is when you get accustomed to it. Now the carding process is intended to attenuate, disentangle the fibres and place them in a more parallel position; also to extract the non-fibrous matter from the genuine cotton, for the better it is *cleaned* and its fibres laid *longitudinal*, the nearer it approaches perfect carding, for that is all we expect from this machine, having made it evener and easier to *elongate* by the drawing roller in the next process. Now to accomplish this carding we have an iron cylinder, 36 in. wide and 36 in. in diameter, with holes drilled in its surface to fasten sheets of card clothing on. These holes are plugged with wood and made secure; these sheets are then nailed on to the cylinder transversely; they are about 3½ inches in width and 36 inches long. These are fastened on with tacks, causing an interstice of one inch between each sheet. The wire in these sheets should be bent at the same angle on the same card. The wire is then subjected to a grinding roller, to sharpen the points, which is done by running the cylinder in an opposite direction to that when it is carding; the *doffer* is speeded up a little, but its motion is not reversed. The grinding roller is then placed so as to touch them both lightly, and runs the same way as *doffer*, but at a less velocity; it is covered with emery. This should not be so fine in its numbers, because it does not get in the wire and grind the points properly, so as to take the cotton and let it go freely without holding it in the body of the wire and choking it into a solid mass like a grindstone, and rolling the cotton on to the doffer in lumps, which you can see by looking underneath before being combed off, and the attendant has had the audacity to say, "that's pretty good for a start." Now, it's my firm belief that a card will do its work best when it is sharp and properly enclosed, without having to *wait* a few days to get in working order. That is all *bosh*; for as you lose your point of wire by working, so you lose the quality of work. It seems ludicrous to entertain such nonsense, yet it is said daily by those you have placed at the head of this department, and will even enter into some fallacious argument in defence of what his Uncle Tom told him thirty years ago, that such was the case after sharpening up the cards. I would like to know about what time or how long it takes to do the work the best, from such an argument? Such sophistry is to be regretted. And let us hope for a more intelligent theory and brighter ideas respecting the grinding and the manner of sharpening the wire so that we may have the best results in saving time and labor and doing the best work.

I have been in mills where the card wire has to be ground every morning for a few minutes so as to enable it to do the work properly; and there has not, up to the present time, been found a better substitute to take its place. Now, if this be true, grinding is a very important act toward success, and should be understood scientifically as well as practically by those who are in charge of the carding, as to how the points can be best obtained and most durable in comparison with the time and labor spent on them. And the knowledge required to do this is not very abstruse, because the requisites are not difficult to obtain. The first is the selection of pure emery, and the second is the numbers or coarseness of it, to suit the numbers of wire required, and the third is the relative speed of the surfaces to be ground, with the speed of the grinding roller and this is not invariable with the same

wire for having ground a cylinder and doffer down to a
smooth surface the grinding roller may finish what has
been partially done, by increasing its lateral motion so as
to bring their points more to a needle point, and all the
surfaces to be ground should revolve at their normal speed
whilst being ground in order that the wire retain its proper
position in both cases and do not lay too hard on with your
grinder, or you may make a point like the edge of a soft
tempered knife which will be ragged, and not let the fibres
off but get all felted in the wires, when such a thing hap-
pens the best plan would be to run the wires into each
other for a while or what is called *facing* them and grind
over again with a little more care.   Now if the carding is
done according to the schedule for single carding, the
grinding may be done every *two* weeks a little to secure
sharpness, for this is absolutely necessary and must be kept
so, no matter how often repeated, setting up your *flats*, as
close to cylinder and presenting as much of their surface
as possible to cylinder, without injuring their points.   I
would invariably use the roller when grinding, although
the *strickle* must not be entirely dispensed with, for there
are times when such an article will do for a make shift and
that is all; the speed of grind roller for cylinder and doffer
is generally about 458 feet per minute and the size of
Emery No. 5's or 6's, make a very good roller.   The size
of wire used varies according to quality and quantity of
work, but from 30's up to 32's and 400 to 500 points per
square inch.

The doffer is a cylinder in form made of iron, with holes
drilled on the face one-half inch from edge, these holes are
plugged with wood and made secure; the diameters of these
doffers are generally 15 inches bare, but with wire on they
are 16 inches in diameter and 36 inches wide.

The card clothing used on these are called fillet of about $1\frac{1}{2}$ inches wide and in one length to suit; this would require $\frac{36}{15}$ equal 24 by 15 by $\frac{3}{8}$ by 3.1416 equal $\frac{1159}{12}$ equal 96 feet long of fillet will cover it. The numbers of wire used are usually one number finer than those of cylinder; in putting this fillet on, the end is made taper, so as to conform to a screw of $1\frac{1}{2}$ inches pitch which it makes by being wound on the surface of the doffer so taut as to almost break it, giving it an appearance of a solid mass of points, which are subjected to a proper grinding.

The surface velocity of this doffer is according to schedule to be 420 inches per minute, and the cylinder will be 1250 feet per minute, with the doffer comb at 420 by 1.3 equal 546 strokes per minute of 1.05 inches length and the draught must be 112 for the weight of lap before mentioned to make one yard of sliver 40 grains in weight, this machine is termed a flat card, with automatic stripper, there are two substantial arches made of iron which fastens on to the iron frame sides, they are set concentric with cylinder and hold the whole paraphernalia, requisite to resist the action of cylinder while carding the cotton. On the top of these arches rest the *flats*, these are made of nicely seasoned white pine, on which the wire clothing is tacked, in an opposite direction to cylinder and doffer ; under the ends of these flats are iron shields to rest on the screws, and help to prolong the wear and exact setting of these flats. One of these flats resting on four screws, two at each end, which are intended to bring the surface of the wire on flat parallel with cylinder surface, these sett screws are made to fit snug in the arch to prevent ever moving of themselves through vibration ; between these is a stud pin made fast in the arches, which act as guides by passing through a hole in the *flat*, always keeping them in a radial and parallel position with the cylinder, there being from 16 to 20 of these flats, and the cylinder having a continuous action on the fibres of cotton which are held by them, enabling it to pull and lay them in a state approaching parallelism ; on this account it is preferable for long cotton and making a sliver that is easier to draw by the fluted rollers, from the fact of it being already laid closer and solid by its fibres being parallel and longitudinal. I have also noticed that sheets made of other material than leather, leaves the web from doffer in holes or in a porous condition, whereas it should be left in one continuous fleece, if properly carded, and there are no kinds of cotton grown, but this make of card will manipulate and reduce to a perfect fleece, or web of cotton carding in better condition than any other card extant up to the present time, it excels in quality, but not in quantity, as the roller card comes next, and first in quantity according to quality, for if we attempt to draw in a mass of cotton by the feed roller, on a *flat card*, you can imagine how the *flats* will be surcharged, and they not being able to retain and allow it to be teased out on account of there not being wire enough to hold this large supply ; for the cylinder is to a certain extent limited in its speed, '' hence,'' the folly of trying to crowd too much work on this kind of a card ; it is necessary sometimes to speed up the cylinder, when you get a grade of cotton, that has a long and strong staple, as it requires a greater velocity to give it the due amount of carding in the same time, but this change is not so readily done as imagined, nor yet is it proper to do it on the *flat cards*, on account of the increased velocity, by centrifugal force, throwing the fibres too forcibly against the flats and choking them up, making altogether too much waste, so you see we would lose by the operation,

so instead we will slower the speed of doffer, and curtail
the length delivered, and by this means we shall improve the carding at the expense of loss in production
there are two things here of great importance, quantity
and quality, and when one is required more than
the other, it must be inversely to each other, but if both
properties are essential, then it involves more cards whether
you do it by single or double carding. I also believe, when
the cylinder takes the cotton from the feed rollers we have
better carding, because of its being held in a mass causing
a greater tenacity of the fibres to be drawn out by the
cylinder, although the wire gets more hard usage than
when it takes it from lickerin. Yet the improved quality
of the carding is more in favor than the damaging effects
on the wire amounts to, for it is on this principle that the
science of drawing cotton is carried out; by the greatest
velocity of surface, doing the teasing out of the fibres,
securing them almost individually by the slow motion of
feed rollers in contact with the cylinder or drawing rollers,
it will be seen at once the difference between the cylinder
and lickerin, for the latter has only half the surface velocity,
consequently it is taken in by lumps and can not be held
on its surface, but is received as such by the cylinder, followed by a deficiency at the expense of the lumps referred
to. This is a consecutive course, which the carding is
subjected right along, showing itself in clouds as the web
leaves the doffer and but for the doubling that follows there
would be an evil that could not be corrected only by being
spun on the mule and stretching it out.

Now these flats have to be ground on a grinder, a machine made for that purpose: it is a cylinder or roller
covered with emery, which has a constant rotary motion ;
some are oscillating, others are traversing, over which these

flats are held in a frame having an alternate motion, moving in a tangential direction over the centre of grinding roller to insure a plane surface, they are then placed on the arches of the card according to their numbers and variety of wire nailed on them; those having the coarsest wire being the first to take the cotton. They are all set close to cylinder, so are the feed rollers and the doffer too, but just to escape contact. You will then adjust the comb just to escape striking the wire on the doffer, and its normal position giving the sliver its proper tension, as this is altered by rising or lowering the comb blade, and in collecting the web together before going through the calender rolls, I would recommend a Holland guide, and when a shield underneath is necessary, it should be set a couple of inches from the doffer, so that the leaves and seed husks will drop between on the floor.

The calender rolls must not have any more surface velocity than the doffer has, because the thickness of sliver will create tension enough to carry it up to the calender rolls.

We have now got all the relative speeds from the cylinder, which will lead us on for a trial to prove by this inductive principle; whether we can make good carding or not, as the web is being combed from the doffer, we must look for a full and clear fleece, taking care the edges are regular and smooth, for it spoils good carding to have ragged edges which can be prevented, by having the wire on the Flats exceed in width the cylinder wire.

We will now suppose the sliver to be leaving the calender rolls in good shape, it will now be an economical choice and not of taste, which kind of process, whether that of coiling it in cans at each card, or by carrying the slivers by a belt up to a railway head; I would prefer the

latter for its doublings and attenuating principles in com-
bination with the cards, if there be not too many cards in
a section

We shall now refer to the schedule which says that the
total weight of all the slivers shall approximate to 360
grains, then according to that our card slivers we called
forty grains to one yard, then $\frac{360}{40}=9$ cards to one section,
which are sufficient for one railway head, these number of
ends 9 are carried on an endless belt, up to the railway
head, forming one large sliver, being collected and laid of
equal thickness before going through the rollers, and from
the weight of sliver, we have decided on making on the
card, will depend the weight of railway head sliver by the
draught, for these are to be nearly constant.

In order to make this single carding good and strong,
we must dispense with the screens underneath the cylinder,
to let the rubbish and short fibres drop on the floor, for
these are injurious to the strength of the yarn, there may
seem to be a loss by such a whim, which would assume
large proportions monthly ; but, the policy of this under-
taking is to insure success, showing by the clamorous custo-
mers the demand for your yarns, which more than counter-
balances the small percentage of waste incurred, and even
this is redeemed by the extra production, but the greatest
of all is the contentment of mind, both at your mill and
in the market, all hands meet you with a pleasant smile,
and who would toil under disadvantages when the above
precept will secure comfort.

The connection between doffer and the feed rollers by
the train of wheels, should be adjusted so that there
will be no jirks or irregular motion of the latter,
seeing that no cotton accumulates round their ends,
to the detriment of the grip in holding the lap, and

securing a uniform pressure obtained by the short levers and weight placed over their journals, which should be equal and permanent when the precise leverage has been determined on from the diameters of rollers and thickness of lap, and prevention of lumps being plucked in by the cylinder, which often is a trouble and serious evil. But this, to a certain extent, is due to irregular made laps, and must be remedied at once ; and before leaving this flat card, let me induce you to have the slivers arranged, side by side, on the belt in the railway trough, to insure an equal pressure by the top rollers of the railway head ; this being so, will require less leverage on them, although it is necessary sometimes to have these slivers more compact, so as to condense and hold the fibres. This is very often the case in dyed cotton, which makes it difficult to draw, and will require more leverage on the rollers ; but use no more than just enough, or your rollers will soon be destroyed. I may remark here that the flat card would not be so profitable, requiring more attention and labor to keep in condition for the carding of colored cottons, which are not so easily disentangled as white cottons are.

THE UNDER FLAT CARD.

With due respect to the inventor of this card and those who are interested in the benefits from it, I feel deeply affected for fear I should give some offence, by describing to you my experience with them, having no desire to condemn or denounce that which, we hope, will ultimately be improved, like all new machines are subject to in their development. I now seek your sympathy in such a task in giving to you my views held from experience ; and it is necessary, for the welfare of all and the machine, to point out such parts that may or may not be improved, as well as giving its estimable qualities, which is not required from me, but from those who have them in use all over the country, will testify.

Now, this card being a more recent innovation, and one that has been a success by the demand for them, this popularity has extended far and near in such a short period, that they have not had a thorough investigation of its merits claimed over their predecessors. There are some doubts about these novelties being accessory to improvement, when properly understood, for their construction seems to be astray from carding principles, one being its manner of presenting the cotton to the cylinder from the lap with the *two* rollers. This application seems to be inevitable from the construction of card more than having any carding virtues; we must not retrograde knowingly, for we want advancement in ideas and curtailing of expenses which have not been considered, but have followed the caprice of enthusiasm by the maddening effect of competition in the market, of who can do most, will get most, has been ringing in the ears of manufacturers. *The under flat card* which has been appreciated for its quantity of work over the ordinary flat card, which gives it an attractive advertisement, causing anxious inquiries by those who are desirous of making a change, would like to see and judge by its results how far they can venture more capital because others have done so.

The card has not shown itself yet to be one of great advantage or capacity, not a desideratum by any means; we have yet to learn how they get such strong yarn these cards are said to make, when from test and experience *it* shows there is an excess of short fibres over an ordinary *flat* card more in proportion than the extra weight turned

off, which exhibits itself when it comes to be twisted and
is as good a *test* as any combing would be, if *such* could
be ascertained with middling cotton from such a card ; its
defects will be shown by the former twisting in its appearance on the bobbin, like a mixing of too much fly and
waste exhibits and the rollers requires to be set closer, also
a greater amount of twist, and by the latter plan of combing, we should get the exact amount of short fibres, showing by comparison an excess, it being a well-known fact
that they reduce the strength of yarn and should be thrown
out along with the leaf and other rubbish ; how is it to get
out when the flats are right under the line of direction and
under the centre line, covering ⅔ of surface of cylinder ?
But these are not the worst features that card has a tendency
to do, and does it without cessation, and that is, by the
rapid revolutions of cylinder the fibres of cotton are lashed
into the underflats, by an *additional* force of gravity, besides the sand and dirt they hold, which should be dropped
on the floor, helps to surcharge the underflats, which is
instrumental in making bad work, along with the extra
labor required in setting them, there being difficulty in
securing a person to do the work faithfully, for you know
how dilatory they are in going about tedious and exact
setting, unless they are paid a premium, but let them
neglect them ever so slightly, and you will soon
discover to your sorrow the evil of having these flats
any lower than the feed roller or centre line, for
it is evident from the remarks previously stated that
there must be an excess of mutilation which short-
ens the much wanted fibres, this evil is not to be
detected so well until it comes to be twisted ; it will all
look very well while the card is just newly set and sharp.
I will remark here an instance which proved to me how a

person can be deceived by what seems to be first-class carding and drawing, which I happened to see in one of our large manufacturing establishments (this has no reference to underflat card, but the *mixing* of short cotton,) being favoured by a permit and escort to go through every department ; after being through the card-room, I was taken down to the mixing room, and there I saw cotton that I thought was not fit to use even for No. 6's, let alone for finer numbers, although the mixing got a percentage of it. I was then taken to the ring spinning and bobbin winding room. I there picked up a bobbin, to examine the yarn, which was and might well be termed *fearful.* I could hardly believe it was all alike, so I got another one, and there was the same irregularity of the yarn. I was struck with astonishment, and it made me think seriously of. the matter, and the conclusion I arrived at, was that it was caused by the diversity in length of fibre and weakness ; and now I will leave you to form your own judgment in this argument, and in addition to this look at the strippings from this card, and this you dare not do perfectly because the double stripping requires too much power, and would break the strippers constantly, which are costly to repair, taking everything into consideration with the general appearance of the card when working, it is untidy all around by stripping, etc., requiring more help to keep the room looking anything like neatness, leaving nothing attractive or desirable in such novelties.

If utility and progress are the ideas of the age let us not have them stunted by such a speculation in that which is the reverse ; we must let the mind go free and investigate the truth, for the benefit of those who have not the courage or desire to search into these things which are doubtful, and might destroy that which is desirable in man, to arrive at perfection in all things. It is this view that I have taken in reviewing this *under-flat* card to speak the truth in every respect; from the conviction arrived at by a diligent watch over its workings, and not from hearsay have I ventured a single assertion, but give my experience with them. For it has been a benefit to the inventor, to the maker, and the machinist in giving them employment, from which I hope they have all derived a reward, and will ultimately improve it, and make it a success. It has not been a sordid or inviduous attempt on my part toward the projectors, but really feel it my duty to say what I have done for the purpose of showing the manufacturers the necessity for such a class of machines, that will improve and earn them a name in the worlds market for making the best yarns, and hold ourselves as such against all competitors, of which there are many; let us endeavor to improve our machines by making them simple by dispensing with, instead of adding more to, so that they will require less expense and labor, striving to excel in the quality of its production, at the same time showing that our ideas are progressive, having a more intellectual desire in our reform, comprehending that our changes shall blend the two elements together, and by so doing we are aiming at more than an ordinary novelty in our improvements retaining our prestige for genius and securing us a reward by the approbation which mankind will lavish on our inventive talent.

THE REVOLVING TOP FLAT CARD.

We have a revolving top-flat card, which does very good work and plenty of it, but its care and attention requires too much skill and labor to make it a desirable machine.

No. 2 TABLE

—FOR A—

40-Inch Roller Card.—The Cylinder 1500 feet per Minute.

DRAUGHT OF CARD 90.

Nos.	Hank Sliver.	Grains Per Yard.	Weight of Lap.	Vel. of Doffer.	Lbs. per Day.
4	.122	68	14.7	1107	180
6	.126	66	14.25	1065	166
7	.128	65	14.	1045	163
8	.130	64	13.8	1034	158
10	.134	62	13.38	1000	150
11	.136	61	13.16	985	144
12	.139	60	12.96	970	140
13	.141	59	12.7	955	135
14	.143	58	12.5	940	130
15	.146	57	12.3	925	125
16	.149	56	12.	910	120
17	.151	55	11.87	890	116
18	.154	54	11.65	870	113
19	.157	53	11.44	855	108
20	.160	52	11.22	840	104

DRAUGHT OF CARD 112.

Nos.	Hank Sliver.	Grains Per Yard.	Weight of Lap.	Vel. of Doffer.	Lbs. per Day.
21	.163	51	13.7	830	100
22	.166	50	13.44	805	96
23	.170	49	13.16	790	92
24	.173	48	12.9	775	90
25	.177	47	12.6	760	85
26	.181	46	12.3	740	80
27	.185	45	12.	725	77
28	.189	44	11.82	710	74
29	.194	43	11.51	690	72
30	.198	42	11.29	675	68

ROLLER CARD.

The roller card which is the best for dyed cottons, and
for quantity of work, will excel all others of which there are
a great variety and choice, each of them having some particular claim, for the construction they have put upon them, and
originality of design by which they intend should improve
the carding, or displace some of the more primative methods
by its obvious and pecuniary merits, which are deserving
of particular attention, and should be encouraged in their
efforts in trying to get at greater perfection of work, than
a mere novelty of *gim-cracks* to keep in repair, of which I
am sorry to say there has been too many recently, in fact,
to superfluity and would bring any manufacturer to ruin, if
he would go to the expense of testing them by experiment,
which always costs money and loss of time, when they are
proved to be of more humbug than real service, and this
leaves you to repent.   We are apt to get confused by the
many recommendations presented to us from individual
manufacturers and large corporations who have endorsed
by their signatures, and which these solicitous drummers
are so ready of presentment to sanction what they repre-
sent, and the anxious desire to get an order for their machines and auxilliary appliances too numerous to mention.
Now this roller card has instead of *flats* what are called
workers and *strippers*, made in roller form and of iron, on
which fillet is wound around in a spiral manner, like the
doffer before described, their axles extending through and
resting on bearings attached to the arches, and by the same
can be adjusted by screws, bringing their surface in contact with cylinder and each other, and by having different velocities the cotton fibres are received and given out to be
teased and torn asunder as the quality of stock requires, and

advancing toward the doffer as they become loosened, and
laid parallel by the action of the reciprocating rollers, in
opposition to the greater force which the cylinder meets
them, carrying with it the already prepared fibres which are
stripped off by the doffer; the cards are usually supplied
with a *lickerin* for the purpose of assisting the cylinder,
which would have too much labor by so great a quantity
of cotton being presented and put through the card, although this licker-in is objectionable in other respects (before mentioned), for it is at this place where most of the
carding could be done, if properly held, of which the lickerin has no claim; it has one; though of assisting in the removal of any dirt or hard knots which the beaters or
scutchers have failed in doing, it being placed in such a
favorable position, and rotating suitably for performing
such work, and not until a better plan is brought out, can
this licker-in be dispensed with altogether, the removal of
moats might be accomplished by placing over the feed roller a
few grate bars that can be set at any angle or distance, to
suit the tangential force the cylinder may have which
will be according to velocity of cylinder, this appliance
would have to be carefully enclosed with a box to receive
all the trash, and that would be almost all there was in the
cotton, for this is the most available place on the card
where such an appliance could be used. There is a roller
and sometimes two, called dirt rollers, fixed over the licker-in to answer this purpose, but they do not do the work
effectually, their wires incline against those of the cylinder
driving the leaf and shells into these from the cotton;
these rollers revolve the same way as the cylinder, and
their velocity like a worker, and sometimes slower; there
is and has been a great many *gim-cracks* for this purpose,
among these, also, may be included one of stripping un-

derneath with two rollers. I think a fancy would be more valuable in their stead underneath, and would help to keep the cylinder sharp and clean if properly applied; this idea of stripping underneath and being stripped in return, and then let go in again is a poor recommendation for making good yarns; there is a point of economy shown by it, but its worth is false, by making that which would have been good, is worthless. It will be in dull times that we shall see all these evils coming against us, and those that have guarded against them will be safe at all times, and demand their price regardless of the market. For our standard to quality must not be interfered with, nor depreciated by these everlasting novelties, unless they insure a guarantee of improvement of facilitating or economizing the same. Now the wire on this licker-in, cylinder, clearers and dirt rollers should be nailed on the same way as they revolve, the others all incline reverse to their revolving motion. The speeds of these rollers are generally calculated and made constant with the cylinder by the machine maker, but this speed is not to be invariably so. Yet they are seldom ever changed. The cylinder on this kind of card may have a greater surface velocity than the flat card has, o that it may lash the fibres into the wire of those opposed to it ; exercising a greater amount of teasing and loosening of the fibres, when assisted by the doffer running slow and the workers moving fast, but this speed of workers and doffers should be inversely proportional to numbers of yarn, that is the higher the count the slower the speed of them both, and a moderate draught that will assist in keeping the cylinder clean, so that the fibres may be saved by being elongated, and getting in the spaces of wire, instead of laying on the surface to be tortured, and made into fly for want of room amongst the teeth, which can all be avoided by careful watching, in order to get the required velocities which are needed on the stock you are using preventing wire from being all choked up.

From the schedule for this roller card the relative speeds are for a forty inch cylinder, 1500 feet per minute, and, the doffer 840 inches per minute, having 90 of a draught with a 11.22 oz. lap to one yard, making a sliver .16 hank, or 52 grains to one yard; these are especially for No. 20's yarn only, and the surface velocity of workers is ⅓ of the doffer, and the lickerin is one-half of the cylinder, the strippers is ¼ of cylinder, and if a fancy is used underneath it might run 1700 feet per minute.

We have found by experience if the cylinder as a greater velocity than the above, it gives out more fibres to be tossed in a turbulent manner over and around the rollers under the enclosed cover, which ultimately collects and forms in rolls and balls, at last dropping by their own gravity on to the rollers, and going through in that condition, making a cloud in the web, and often breaking it down, causing a thick place in the sliver which will show itself wnen made into any kind of a fabric, but it is often detected before being spun, for they cause a good deal of waste to be made by breaking down in the various machines it has to pass through, being a source of trouble all the time unless pulled out and thrown in the waste can, to be worked up over again.

I would prefer doffers of large diameters, in order to have a less number of revolutions, and because of there being a greater surface presented to cylinder, taking more fibres with it when clearing the cylinder of its work, which will prevent in a measure too much fly and waste, and the cylinder from getting chocked up. I hold the same opinion regarding the workers, because the fibres of cotton will be lashed in more; it is also important to have the cylinder covered with filletting instead of sheets, causing less commotion of air under cover, for the sheets act as a fan when the cylinder has increased speed, we are also benefited by getting more wire on its surface by dispensing with the interstices between the sheets.

But for a flat card, I don't think fillet is so good as sheets on account of there being no revolving action with flats, opposing the cylinder surface to start or raise the continuous fleece of cotton, which will accumulate and make neps and form in rolls for the want of something to start it like a fancy does, so it is obvious here the preferment to sheets over filletting, by the interstices which cause a start and allow the doffer to take its compliment of fibres regularly, and in ratio with the feed rollers after being attenuated to the extent of the draught.

The card will do 104 pounds per day of ten hours on this particular number of counts that is 20's yarn, and show greater strength of sliver with less condensing than any other card will, saving waste and facilitating its progress up to a certain point, and then the flat card supercedes it, because its fibres are laid more parallel and condensed, and making it more favorable for attenuating to firmness and producing a higher number of counts, everything else being equal, yet I believe up to these counts of yarn, the roller card would be found to have the most economy in it, and produce the best results, taking it altogether.

We will compare their production, and show the difference between the flat card and roller card: Flat card, doffer velocity, 420 inches per minute. Roller card, doffer velocity, 840 inches per minute; the sliver of flat card weighing 40 grains equals .208 hank; the sliver of roller card weighing 52 grains, equals .16 hank. The number of inches divided by 50, will equal number of hanks for 10 hours, and number of hanks divided by hank sliver, equal number of pounds for 10 hours. Flat card, 420 in. divided by 50, equal 8.4 hanks divided by .208, equal 40 pounds. Roller card, 840 in. divided by 50, equal 16.6 hanks divided by .16, equal 104 pounds per day. The ratio being 104 divided by 40, equal 2.6 flat cards for one roller card, this consuming $\frac{1}{2}$ horse-power and the flat card, .25 horse-power, then $\frac{2.6}{4}$ equal .65 horse-power, so the difference in favor of roller card would be .65 minus .5 equal .15 per cent. in power, according to equal quantity in weight in pounds per day.

The investment will be less ; there will be a small percentage in loss of time for stripping and cleaning, the cost of labor and incidental expenses, conjointly, will be more on the whole than the flat card will be ; its favor will only extend up to No. 20's yarn, for the preparation is not such that will warrant high speeds and larger draughts required for finer counts, but on the whole you can spin the specified numbers cheaper with the roller card than otherways, by the great amount of work which this card can produce by its (*modus operandi*), giving us convincing proof when looking at the web, as it is combed from the rapid surface of doffer, both clean and well-collected, forming a sliver of great strength by the tenacity of its fibres which are the constitutive principles of this card, created by the revolving rollers and their relative velocities, causing a contraction and then a distention of the volume of fibres, which are taken up successively by them from the cylinder and returned again to be repeatedly distended, and straightened out lengthwise amongst the wires, and those which have been driven by the centrifugal force on to the workers and not yet disentangled, revolve round

to be teased out, when they again come in contact with
the wire points of a superior force, tearing and loosening
them to be carried forward along with some that are still
tortuous and short, for you will perceive that the fibres
left floating on the surface of these rollers not already
straightened, will conform to their own whims and become
a little more tortuous than when they are kept under the
scanty space of cylinder and flats, which give them a more
positive distention.

But the point at issue is this—that these short and tortuous fibres that have gone along, help, by a process of linking
with each other towards giving the sliver that strength and
tenacity before mentioned, we must bear in mind that we
are working an inferior cotton, and that the sooner it gets
out of the card the better, if cleaned and loosened enough
for the straightening will be sufficient, for these counts and
class of cottons required, which will not stand such an
amount of carding, as the longer and strong fibre will. I
have already set the doffer velocity at double the speed of
a flat card doffer, when making this same counts, and will
be found not overrated. Knowing at the same time that
the shorter or less the length delivered by the doffer, is
supposed to improve the quality of work or carding, it is
also expensive to do so, and will not permit of it unless
you are remunerated for your loss of time and waste in
the price of your yarn. I think if you will refer at all
times to the schedule, and work as close to it as possible,
you will come out safe, although there are some kinds of
cotton, which might cause a deviation from it, but this
must be a secondary trial. I will give you here the different
velocities from the cylinder and doffer.

The Cylinder goes   1937 in. for 1 in. of feed.
The Lickerin,   968 in. for 1 in. "   "
The Doffer,   90 in. for 1 in. "   "

The Comb 1 1-16 in. 1400 strokes per minute.
The Doffer, 840 in. " "
The Cylinder, 1500 feet " "
" do 64 in. for 1 in. of worker.
" do 21-5 in. for 1 in. of doffer.
" do 4 in. for 1 in. of strippers.
" do 2 in. for 1 in. of lickerin.
" do 1 in. for 1-16 in. of Fancy.

This cylinder is 40 in. x 40 in. and makes 143 revolutions per minute. Where this card is in use, and has two (2) rollers underneath the lickerin, the middle one may be set up to lickerin and cylinder, but the lower one should be covered with fancy wire, being the same numbers, or thickness of wire as the cylinder, but not so closely set in the fillet; this will help the carding materially, and it should have a good clearance to throw its brushings on the floor. I have been thoroughly convinced of the efficiency of this roller, by making two good points, and requires but little attention for the work it performs. I would also recommend good strong diamond-pointed wire for the lickerin, it being the most serviceable for doing such work at that velocity. You will see from the schedule that I have made a special draught to be used for all numbers under 20's, and a special lap to be used for each number of counts.

The speed of doffer varies with the counts also, but not in ratio with the numbers, as is usually done with cards having the coilers attached, which is done usually by proportion, by a change pinion having its motion direct from cylinder; this change being modified from a compound of the weight and length to suit a constant draught, by which the manipulation will better correspond with the counts to be made by an alteration of weight of sliver and lap, and the speed of doffer inversely to counts, and which is entirely too rapid for connecting a *railway* trough and *head*, the delivery being 840 in. per minute from the doffer on these No. 20's, giving us double the length of the surface velocity of *back roller* which runs on the *railway head*; "hence," we are compelled to use the coilers, if the quantity of work must be kept up according to the schedule; so we will make no exception to this, and by dispensing with *railway head*, if the usual *two* drawing heads are not sufficient in getting the sliver even and the fibres parallel, we shall be obliged to resort to an additional drawing head. We must examine this closely and see whether our stock or class of cotton will admit of this extra doubling or not, for it is here that you must decide to sink or swim from the effects at this point, which your preparation will have on the sliver, if it becomes lumpy, its value is irreparable.

RAILWAY HEAD.

Now comes the railway head, an auxilliary machine to the cards, for assisting and reducing the large volume of sliver so formed by a collective number of them going up to the back roller in mass. This roller must have a surface velocity equal to the doffer of flat card—420 in. per minute, in order to keep the tension right. There are four lines of rollers on this *head*, having interstices $1\frac{5}{16}$ in. from fourth to third roller, and $1\frac{5}{16}$ in. from third to second roller, and $1\frac{7}{16}$ in. from second to front roller, with a draught of 3.75 when belt is in centre of cones; then, if the whole weight of slivers from cards is to be 360 grains for all numbers of yarn, we can easily find the number of cards to one section from the weight of card sliver, for No. 20's

is 40 grains, then $\frac{360}{40}$ equals 9 cards to 1 section, arriving
at a system which will commend itself highly by giving us
the proportionate doubling and draughts for the counts to
be manufactured ; here is the sliver from railway head
—$\frac{360}{3.75}$ equals 96 grains to 1 yard. How simple and how
good is this method, for this is a weight that is tenable with
the leather rollers and with ordinary attention will hardly
ever be seen to cut, by having a volume so well proportioned
to its work. This sliver is expected to be made from a
web having smooth selvedges, and the hole in trumpet to
be of an oval shape that will prevent the sliver from being
of a tape like form, and the *condensing* of it must have particular attention, there being two things to be considered,
one is to give it sufficient strength, and the other to allow
the rollers in *drawing frame* to attenuate it without an excessive weight on them. I have given the draught of *railway head* 3.75, but it is reasonable that you should know
how I get at this and not let it be thought a *guess*. No; it
is derived from a *rule*, which answers for four lines of rollers—sq. rt. of 64 multiplied by hank sliver to be made which
is 96 grains equal .087 hank, so 64 by .087 and then extract
the square root will give 2.4 draught for front and second
rollers, and then multiply this 2.4 by 1.56 a constant number for the other *two* (2) draughts, then the product will be
2.4 by 1.56 equal 3.75 the whole draught and as the hank
sliver varies so does the draught between the front and second
rollers, the alteration being altogether with these two rollers coinciding with the action of the evener as the tension
varies in the trumpet, moving the cone belt laterally to
correspond with it and the required draught, so that the
sliver may at all times be one weight of 96 grains, and to
obtain this requires some precision for the instant the tension acts on the trumpet, the *dog* should drop in the ratchet

wheel and not have to keep struggling, as it were, by there
being too much surface on the quadrant extending beyond
the point of dogs, which will cause a difference of pressure
on the trumpet to throw them in gear, leaving the change
in sliver go, before the *evener* has acted at all, such blun-
dering as this is a serious matter, and must be remedied
by filing off these extensions on the quadrant, leaving such
a distance (by the stripping off a light sliver with your
fingers at the back), will allow the dog to drop in the
ratchet, but do not be so precise as to leave the evener always on a move.   There is also a dog on the shaft to which
the trumpet is attached, having two (2) points at right angles
that come in contact with the front of roller beam, and the
distance of these points from it, should only be such as
will give the lever that angular velocity to allow the dog
on the quadrant drop off, in the ratchet, leaving you only
the nature of the atmosphere to regulate against.   This
machine has traces of evil as well as good qualities, there
must be a portion of evil go through before it can be recti-
fied and made good, for the former is the governor and will
evidently show up its imperfections at a disadvantage when
the changes are quick and transient, but other ways, its
good qualities overbalances the evil ones and leaves us yet
in favor of its usefulness and will hang on to it until something better turns up as an improvement.   Now if this
change of tension could be given to *evener* before it reaches
the back roller, we might sanction it a complete machine
and is worth trying to accomplish by its necessity, and
any one would be amply rewarded for his genius, if pro-
tected by patent right.   The rollers of this machine are
generally shell, covered with leathér and weighted at the
ends, having wood saddles, each one straddling (2) rollers ;
over the centre of the interstice, hangs a stirrup on the

saddle, going through a hole in the beam and attached to a lever underneath, connected to that, is the other lever for the other (2) rollers, the whole weight being given by another lever attached, on which hangs the weight, and can be moved into different notches to give the required pressure, which should be about 240 lbs. altogether, on rollers.

The condensing of the sliver is assisted by the calender rolls in union with the narrow slit of trumpet, being careful not to have too much pressure on these calendar rolls, which I previously remarked ; over these leather rollers is placed a wood clearer covered with flannel to keep them clean, and loose fibres from lapping around them, there is also a clearer used underneath the first and second fluted rollers to prevent the loose loose fibres from lapping on them ; this is held up by a counter-weight, fastened to a lace whose other end is fast to clearer, and pulled up close by the velocity of front roller on which the lace rests, and in placing the dish for cotton *can* to stand in, let the inside edge of *can*, when vertical, be under the centre of calender rolls, with a rotation of about 50 feet per minute, this motion ought to be very slow in order to prevent the sliver from twisting so when going under the roller of the next machine, which makes it troublesome at times on account of its volume being so large and turning on its edge, raises it off the other sliver that is under the same boss, and loses its traction going through in chunks.

Under the roller card system we are obliged to use *coilers*, which are attached to the card for the purpose of conveying the sliver into a *can*, as it leaves the doffer this arrangement, is a section of itself and independent of any contingencies that may happen to others beside it, and in this respect it is superior to the other plan. It is on ac-

count of the great length of sliver given out per minute from the doffer that we have to apply such a useful
compact and neat machine, requiring little attention
and packing it in the *can* in such beautiful coils
preserving the fibre from any disturbing element also condensing them and making the sliver in a better condition
for drawing, but there is a little more labor attending to
this system, by having to handle so many *cans*, they generally being of smaller capacity, causing them to run out
and fill up faster, hence the extra labor in removal. Now
this coiler is capable of doing a large amount of work for
which it is adapted, that of packing cotton into the
cans, but it is not required here alone, the card only
having to do about half the quantity, it does at the drawing
frames.

It will be necessary to explain how the *can* must be
placed so that the coiler may fill this *can* to the best
advantage, both in quantity and quality, with as little dis-
turbance of the fibres, as possible. We will call the
diameter of coiler 8 in., and from the centre to the inside
of hole 2¼ in., adding to this ¾ in. for the thickness of
sliver, making the throw 3 in. altogether; after this has been
ascertained, we will fit an inch strip of wood across the plate
the coiler revolves in, and get the centre of coiler here, we
must know the diameter of *can*, we will call it 9 in. inside,
then the radii will be 4½ in., and from this you subtract
the throw of coiler, then 4½ in., minus 3 in. leaves a difference of 1½ in. to be set out from the centre of coiler. We
have already found the centre of coiler on this strip of
wood, then measure off 1½ in. from this centre towards
the front, you will then bore a small hole through the
strip at this point to admit of a string being passed
through, on this is a plumb-bob below, giving you the

exact centre of the *can-dish* which must be set perfectly level in order to give the *can* a true vertical position; these *cans* are generally 3 feet long, but to give them a good clearance when removing them, you must let the dish be 3 feet 1½ in. below the coiler; by doing this, you save the minder a great deal of labor. The coiler must have sixteen revolutions, for the *can*, one, or thereabouts; it is usual for the machine maker to fix that, and it is well to know anyhow for future changes which may arise. This explanation will probably be sufficient for all practical purposes, without going into the abstract of this ingenious machine; it is admired by a majority for its labor-saving propensities, and disregarded by the minority, by having a surplus of mechanical appliances to the detriment of its work, and keeping in repair these superfluous devices, when the work can be done without them, but not so neatly as they do, with such little waste and less expense. It is like most other machines, they can use them best, that have been trained up to them; for it is the help when taking this view of it, that puts in a particular claim on its virtues.

No. 3 TABLE

—FOR A—

COARSE SLUBBER.

Nos.	Hank Sliver.	Grains Per Yard.	Lbs. Per Sp.	Rev. of Roller, 12 x 6	No. of Turns.	Layers.	Draught.
7	.37	22.5	31.5	230	2.8	6.1	2.88
8	.38	21.9	29.48	214	2.82	6.18	2.96
10	.41	20.3	26.25	197	2.94	6.4	3.2
11	.42	20.	24.13	178	2.97	6.5	3.25
12	.43	19.38	23.94	180	3.	6.55	3.35
13	.44	19.	23.	178	3.02	6.62	3.42
14	.46	18.1	22.	173	3.1	6.8	3.6
15	.47	17.7	21.25	170	3.14	6.86	3.67

Nos.	Hank Sliver.	Grains Per Yard.	Lbs. Per Sp.	Rev. of Roller. 10 x 5	No. of Turns.	Layers.	Draught.
16	.47	17.6	20.42	227	3.15	6.86	3.69
17	.48	17.3	19.8	217	3.18	6.91	3.75
18	.49	17.	19.	210	3.2	7.	3.82
19	.5	16.6	18.43	200	3.24	7.08	3.91
20	.51	16.3	18.12	200	3.28	7.15	3.98
21	.51	16.3	17.32	186	3.29	7.16	3.99
22	.52	16.	17.	183	3.3	7.2	4.
23	.53	15.9	16.45	175	3.34	7.3	4.08
24	.53	15.7	16.	170	3.35	7.3	4.14
25	.54	15.4	15.6	167	3.37	7.35	4.22
26	.55	15.15	15.24	163	3.4	7.42	4.28
27	.55	15.13	14.76	154	3.4	7.42	4.29
28	.56	14.8	14.37	152	3.43	7.5	4.38
29	.56	14.6	14.15	150	3.44	7.5	4.45
30	.57	14.4	13.78	146	3.46	7.55	4.5

THE DRAWING FRAME.

The drawing frame is the next machine in order in the manipulation, or what is termed in the preparation of middling counts, it is intended to draw and even the sliver by doubling 2, 4, 6, or 8 ends according to the fineness, and quality of yarn required, and then lengthened out by the roller to any amount desired, this is done by each successive roller revolving faster than the back roller, and producing a length approximating to the number of hank sliver, which is done by the draught. We have a variety in kind and construction, and you will have to decide which of them will best perform this drawing; we have machines with 5 line of rollers, and 4 lines, and 3 lines, their being a draught between every line of rollers. We have also a frame having 4 line of rollers, with only two draughts, that is the fourth and third line have a draught, it is then drawn through a guide made so, as to contract it a little before going through second roller, but has received no extension from third to second until the front roller gives it another extension or draught; these are called double draughts. On the other frame of 4 line rollers, we have three draughts by them, all being set at proper intervals, and making a whole draught. Then there is the 5 line of roller frame, which has three draughts, but similar to the 4 line, having double draughts in its working, and is a first-class drawing frame, and if this style of drawing frame was made so that the 4 line of rollers would start simultaneously, the drawing would be free from cuts in starting and stopping the machine, neither would there be any when it was running, there being only two rollers used for a sectional draught; there is then a space of several inches intervenes before going through the front

section ; in this space the web is collected and drawn
through a traverse guide that condenses it ; there should
be no draught at all in this space, but it is extended by the
front section to the required draught, the product of the
front and back sections making a whole draught, with an
additional fraction by the calender rolls.   The preference
of this method over the other is this: Where there are
several rollers and draughts connected with the whole
draught, the preceding one interferes with the making of
regular drawing, by its being attenuated and spreading on
the rollers as it advances, causing the fibres to loose their
tenacity, and by the interval being so great, causes the
sliver to break, and through its dispersion makes irregular
and cloudy drawing, because in the very act of drawing,
the slivers volume increases at the moment it is suffering
this extension, for the fibres rush out like rays of light, under the sudden action of drawing rollers, freeing themselves
from the compact received by condensation, and if allowed
to advance under another series of rollers, without being
condensed, is simply ruinous to the sliver ; this is the great
objection to a series of rollers, but it can be ameliorated
some by reducing the preceding draughts to a minimum,
and letting the front and second do the maximum, there
is then an excess on these two rollers, and would be done
much better by the double draughts, although this method
is supposed to relieve these exceptions, by the loosening
and better preparing it for the front roller, to execute this
extra work imposed upon it, and may be seen at work in
every cotton mill, there being a majority of these kind
used, and almost universal.   But that don't satisfy and
break up my argument ; the public opinion, and their
choice of these machines, gives no reason why I should
not differ with everybody else, if I have found a more

reliable theory, and practice to back me up, and that which I have before propounded and which I think is a part of the true science of drawing, and on this principle is the art of cotton spinning carried out, for by twisting we can conceive the idea of having the sliver condensed, showing how, by the friction of the fibres, they are held to be drawn out in a manner that is consistent with the principle of systematic drawing, and I think any other theory propounded, will not give such satisfactory results, as the principle already explained, for by properly understanding this theory, as you can easily prove by experiment and will no doubt convince you that this is the key note to strike for making good yarns. Yes, it is a panacea for all its ills, when properly applied. But now we have come to the skill and judgement required in the application, or adjustments, of the *intervals*, and *weights*, and *volumes*, also the nature and state of the cotton to be manipulated so carefully, these are the essential points, and can only be applied by them who are familiar with their workings, to place them in their true positions, so as to produce the desired effect, which is a cloudless and clear sliver, with the fibr s close and straight to each other, brought forth in a lucid like manner. The arranging of the draughts are to be considered also, and this depends how far the cotton has advanced in its riband-like process. For I presume to say, that we should have small draughts to commence with, when the cotton is not fairly *straightened* out, and must be done when it is an untwisted sliver in a careful manner, and by so doing we shall then be at liberty to increase the draughts all along in every progressive machine, having the greatest on the mule or spinning frame. I have endeavored to show you my view of the machine, that would be most suitable to perform this elongation, but you must understand it is in a great measure depending how the functions of this most important machine are arranged and applied. It is not to be inferred, because you paid a big price for it, and was made by such an eminent and celebrated machine maker, that you are possessed of such a one that will insure you perfect drawing, without considering how the functions of this machine are to be adapted to the class of work it has to perform. It is generally supposed that the machine maker has furnished all that is necessary for making the drawing ; and I have found it just as customary, for those who have the charge and running of them, to think it an act of malfeasance for any importune suggestions or casual remarks, as to making some change before you commence operating with it, no matter how beneficial they might be. It is utterly disregarded, for their confidence would be misplaced by so doing, and the reverence they have for this particular machine, giving indisputable organization in all its parts, leaving no alteration on any account whatever to be done. Well, then, if this be so, shall we leave it in a passive manner, and submit to this indisputable claim without investigating this pre-eminent authority and infallible right which they have inspired from some classic acquaintance, whose intimacy is probably of a binding and servile nature for better appointments. Excuse me, and let me expiate a little and say, What boy will not climb to get the cherry, if he does lacerate his body and tear his clothes.

Now, here is a piece of bigotry, which I recently witnessed, that brought out these remarks above, and have frequently met with such persons and their sophistry.

It is not unreasonable to think, by any one acquainted with the manufacture of cotton, that where the number of

counts differ so much in their range, but it will effect the
preparation on account of the change in class of cotton required. I am now speaking of what should be done, and
not what is usually done, that is, ramming all kinds
through without any alteration, except in the number of
hank roving, which makes it compulsory in extreme cases,
and that itself seems to be little understood theoretically.
But we must go farther back than that, and see whether a
class of cotton with a disparity of fibres will, under the
same treatment, bring out the same class of yarns,
as a class of cotton with a uniform staple. This is
what we are driving at; but what astonishes me most
is that the former, under the same gradation as the
latter in preparation, makes an equal claim on the
price of his goods with the latter, in the market, and using an inferior and cheaper cotton. Just
let me say this, that it is impossible for any one to do it,
unless at great expense of getting all the short fibres out,
so as to get a uniform staple, and then it probably fails
for want of strength in fibre. In the treatment of these
would be to get the cotton out of the card quick, for this
inferior kind, and have small draughts which means coarse
numbers, also, to set the rollers closer than otherways,
and reduce the breakage of draughts if it becomes a real
necessity, in order to give cohesion to the sliver, and prevent spreading on the rollers, and will assist the weighting
in a measure, which will require an additional pressure,
the volume of sliver remaining the same. The doublings
ought not to be so great as the other, for they entail large
draughts, nor will they stand equal draughts with
equal doublings repeatedly in the drawing frames,
on account of the disparity of fibres and being in
an untwisted sliver, also, not being able to set the inter-

vals of rollers to suit all lengths of fibres, but as they become more parallel, they release themselves, and slip out better, riding along with them, the short fibres that cannot be reached, and forms in clots which by repeated drawing on this machine, begins to make a lumpy sliver, and would be better before going so far to turn it over to the slubber, and get some twist put in and get a closer bite with the rollers; again, there can be two extra doublings by using an intermediate frame, which would cause a reduction of draughts in the three frames. I only mention this because it coincides with the drawing to be made, and must be understood whether such is to be used, if so by referring to the schedule we find the finished drawing to weigh sixty-five grains to one yard, with five of a draught and four doublings, and the first head of drawing to weigh eighty-five grains sliver, with four doublings, calling the railway head sliver, ninety-six grains to one yard, and constant for all numbers or counts, when making warps from good middling cottons. But if the intermediate is dispensed with, the standard weight of finished drawing will be fifty-two grains to one yard, with five of draught on both heads of drawings, and four doublings on each head, calling the railway head sliver, eighty-four grains to one yard, and constant for all numbers under this preparation, and if using the regular four sliver roller drawing frame, having three draughts, I would propose 1.2 in. for first, and 1.3 in. for second, and 3.2 in. for the front and second rollers, and if the doublings are required to be changed; the front draught will be in proportion to the hank sliver, but keeping the breakage draughts at 1.56 in. I must here give you the *rule* for getting this draught, sq. rt. of 64 multiplied by hank sliver, then multiplied by 1.56 in. equals the whole draught. The intervals being $1\frac{3}{16}$ in., $1\frac{5}{16}$ in., and $1\frac{3}{4}$ in.

for front and second; but if you should happen to have
a three line roller drawing frame, then the *rule* is sq. rt. of
128, multiplied by hank sliver, then multiplied by 1.11 in.,
equals the whole draught; this 1.11 in. means the breakage
draught for three roller frames, and the intervals, 1⅜ in.
each, and the weighting of top rollers to be from 50 to 60
pounds on each roller having two bosses, but if using only
one boss, a little lighter weight might do; however, you
can judge for yourself after giving this a trial, for an excess
of weight ruins the leather rollers and should be avoided,
but not to endanger the drawing. I shall give you another
method here, of arranging the drawing for ordinary cottons, which works well, and may say, admirable. Here
the railway head draught is 3.3, and the sliver weighs 100
grains to one yard; you only put two of these ends up at
the first head of drawing, with a draught of 3.75)100 multiplied by 2(equals 53 grains per yard. Then we will put four
ends up at the second head of drawing, with a draught of
four, making it 52 grains to one yard, which weight cor-
responds with the other plan. This method having the
principle of drawing in its favor, but not the doubling,
which must be governed by the class of cotton. If using
ordinary's, then this method will do, but for good midd's
cotton I would prefer the extra doublings and draughts. I
must impress on your mind here the importance of the
draughts, which are not to be considered from the number
of doublings but from the volume of cotton passing under
the top rollers, and must be in the proportion of hank
sliver being made. For instance, if four ends are put up
weighing 50 grains each, equal 200 grains, we must not
infer that if two ends are put up weighing 100 grains each,
equals 200 grains also, that we are to change their draughts
on account of their doublings.

The production of these drawings from their respective
weights of sliver, and velocity of front roller at 144 ft. per
minute, will give us 216 pounds per ten hours. When
making a 65-grain sliver and when making a 52-grain
sliver, we shall get 175 pounds per day on each delivery,
making a .13 hank and .16 hank respectively. I generally
set the rollers at first, making the back roller 4 inches from
front, that is, from centre to centre, it is then an easy
matter to adjust the other two lines to suit the working of
the material, and for the protection of the top leather
rollers, it is preferable to have about 12 flutes to the circu-
lar inch, and should be made so that the flutes will be
irregular, to prevent the top roller forming a surface corresponding to the bottom roller, by the heavy pressure
required to attenuate such large slivers, for if the leather
becomes corrugated the damage to the drawing becomes
inevitable, and must be replaced by smooth-surfaced ones
immediately they are discovered, having always a reserve
on hand sufficient to cover these emergencies, and if these
shell rollers were made to suit the width of the doublings, we
should not require any traverse, which is an evil on a drawing frame, we could then make our rollers shorter, which
would be an advantage in the weighting, and holding the
sliver more effectual. We should have four ends brought
into one at the trumpet, then, with a less angle, this being
repulsive, and the trumpet should have a bore of $\frac{3}{16}$ of an
inch, and let the calender be lever weighted, so that the
sliver can be condensed to suit circumstances, taking care
to have no draught between them and the fluted rollers,
only what the thickness of sliver would create, we must
also have the guides to the back rollers as close up as possible, so as to keep the slivers close together, and prevent
thin places underneath the leather roller, this being a link

in the chain of measure for good drawing, and for economy would use coilers on the first head, but on the second or finisher, it is necessary to see, that the drawing is made perfect, and kept untrammeled by any friction or excess of machinery that may injure it unseen, and bring us trouble that would be very expensive to remedy, therefore we will dispense with coilers, in this head of drawing, and let the sliver drop into ten in. cans, they revolving reciprocally enabeling the sliver to leave the can without twisting it so, as it leaves the can, they will also have to be packed by pressing with your hand carefully, so that the fibres may be kept undisturbed, using every possible means of prevention, to secure the parallelism of these fine filaments, before becoming twisted, the necessity of this is explained as it progresses. The utility of this ingenious mode of nipping hold as it were, with the points of your fingers, and drawing out the fibres longitudinal, by their infinite number and small points, in such an orderly and mechanical manner, being a decided improvement over the master-stroke of the points of card wire, which of themselves cannot nip hold of these innumerable points, but as they present themselves they are hooked off by their loops and spiral contortions, and carried into the meshes of wire to be lashed unravelled and tore, and straightened partially according to the capricious nature of the fibres, which are thus combed off indifferently compared with the masterly manner of these drawing rollers, it is as much superior to the card for drawing, as the comber is for cleaning. Thus showing how progressively arranged are the different manipulations which the cotton has to undergo, as it advances in its career towards the attainment of a perfect thread.

And yet there are *theories*, visions and ideas advanced, that this machine is superfluous, and a complete deadhead,

having it is said, been proved by microscopic investiga-
tions, that the parallelism of the fibres are in better condition, before going through the drawing frame, and only
becomes necessary here to reduce the sliver to a certain
weight, and in performing this, it produces an uneven
sliver, which can, be easily done if placed under such in-
experienced treatment.   Now such a statement as this,
and to have come from a scientific investigation, whose
authority runs pretty high among those persistant in-
quirers where it originated, and they have had the audacity
to bring before the public such startling results of their in-
vestigations, which is calculated to do more injury than
benefit the manufacturer, who is always ready to take advantage of anything new that will aid him in doing his
work in a more economical manner, although his former
experience may have been extensive, having acquired a
thorough knowledge of how this machine does its work, and
with such precision, knowing that it is an indispensable
machine, so far as he has proved the requirements necessary for making good yarns.   He knows if this machine is neglected, and should in any way get out of order,
that his yarn becomes worthless, if it does not perform its
functions by drawing and placing the fibres more parallel,
and making it evener by the doublings than when it left
the cards.   Yet, in an idle moment, he picks up a pam-
phlet to read a series of experiments, which are undergoing
a severe test, by some of these scientific men, who pro-
fesses to have discovered, and by their machinations to
lead the public to believe by some new evolution, he can
dispense with this useful machine, and he being a contem-
porary with Darwin, leaves it questionable with this in-
quirer whether such an appendage, is not more of a nuisance
than being useful.   However, after reading the flowery

statements at which these experiments brought forth, or
as represented, he becomes sanguine of its merits, and
rushes headlong into this artful device, he arranges his
carding and railway head sliver to conform with the
draught of the slubber, and through the excitement everything seems to run along satisfactory, until he comes to
watch the ring spinning, here his heart is bowed down,
and wonders why it does not run so well here; after re-
flecting deliberately, he thinks by altering so and so, will
help it; after a while, something else suggests itself, and
there he goes making all these changes, and still no better.
At last he reconciles himself by thinking: oh, well, that is
good enough, I have seen worse than that, and maybe it
will come out all right after a while. He again returns
to the card-room, and is delighted with the change, having
dispensed with the drawing frame, and the help required
to attend them, he could never think of going back to the
old system. "No, that is too absurd altogether; I am a
little ahead of my neighbors and competitors now; they
are too slow in picking up these novelties, instead of taking advantage of them right away. Don't talk to me
about changing back again to be laughed at. No, sir." In
a week or so the spooler complains, and the warper, and then
the weaver who says it is impossible to weave it. It is now
about time for one of these scientific men to bring along
with him, a micrometre, and give us a little of his measur-
ing experiments, showing us how many different numbers
of yarn he can claim in 36 in. length more than by the
old system, which I will warrant is double, if made into
medium counts, but if very coarse, then the unevenness is
not so easily seen, for it is when you begin to extend the
draughts that the evil commences, by the reason of its
fibres becoming twisted before they were laid parallel, and

I may state here, affirming, that it is impossible to draw
and keep an even thread from such a preparation, and I
mean this as an admonition to those who have not had
this experience, will benefit by it, and leave such a course
alone.   In reviewing the course of this drawing frame, it
will be seen that the doublings are few, only being sixteen
when using the railway head, and making middling counts;
but for coarser numbers using the worker and stripper card,
sixteen doubling will be sufficient, but if higher numbers
are required, another drawing frame should be used, in the
preparation using a more equable staple of cotton, if it is
necessary for the articles from which your yarns are intended to make, whether it is to be a close, smooth and
wiry thread, or a loose, bulky, fuzzy thread, this all depends on the doublings and parallelism of the fibres, and
this drawing should be done very cautiously ; the amount
of draughts used are to be compared with the state of the
fibres, and will not admit of large draughts while in a
confused state, and should never be attempted until the
preparation is complete, and a portion of twist in it, and
there the extension may be carried out according to
schedule ; but the longer, and finer, and stronger the
staple, so may the doublings and draughts be increased
when spinning fine numbers, hoping these few hints may
be productive of making *good drawing*.

No. 4 TABLE,

——FOR A——

FINE SLUBBER.

Nos.	Hank Sliver.	Grains Per Yard.	Lbs. Per Sp.	Rev. of Roller.	No. of Turns.	Layers.	Draught.
4	.417	20.	22.5	238	2.96	6.45	2.6
6	.5	16.6	19.96	230	3.24	7.08	3.13
7	.54	15.3	18.6	221	3.37	7.36	3.4
8	.56	15.	17.53	204	3.41	7.5	3.46
10	.63	13.15	15.53	192	3.63	7.95	3.95
11	.64	13.05	14.33	172	3.67	8.	3.98
12	.67	12.25	14.3	176	3.75	8.2	4.24
13	.7	12.	13.66	174	3.84	8.4	4.33
14	.73	11.25	13.08	172	3.92	8.55	4.62
15	.74	11.15	12.63	166	3.95	8.62	4.66
16	.76	11.	12.18	156	4.	8.72	4.72
17	.78	10.68	11.66	157	4.1	8.85	5.02
18	.81	10.28	11.33	156	4.12	9.	5.10
19	.83	10.05	10.95	153	4.17	9.1	5.17
20	.85	9.8	10.73	153	4.23	9.22	5.3
21	.87	9.6	10.3	149	4.28	9.35	5.41
22	.89	9.45	10.1	149	4.33	9.46	5.5
23	.90	9.26	9.783	145	4.34	9.5	5.56
24	.92	9.05	9.5	142	4.36	9.61	5.71
25	.94	8.9	9.276	140	4.4	9.72	5.84
26	.95	8.8	9.03	137	4.46	9.78	5.97
27	.97	8.6	8.783	136	4.5	9.85	6.
28	.98	8.5	8.56	134	4.55	9.9	6.11
29	1.	8.3	8.4	133	4.6	10.	6.26
30	1.02	8.16	8.16	130	4.65	10.	6.45

THE SLUBBER.

The *slubber* is the next machine which takes and uses the drawing, and puts the first twist in, and is then called slubber roping; these machines have three lines of fluted rollers, and a number of spindles and *flyers*, varying according to the amount of work required, they are generally made from 40 to 80 spindles, whose length is also determined by this amount; there being *one drawing sliver* for each spindle, taken from the can at the back of frame, and carried over a roller which assists in lifting and keeping the sliver straight, it being brought out over the can to relieve the friction on the edges. This *roller*, conveys the sliver to the guide close to the back roller, under which it is drawn, this sliver has been carried up here by a motion having the same velocity as the back roller, to prevent it being torn and its fibres displaced and this important feature will be assisted by having the cans made of such a *diameter*, that will be equal to half the length of *roller box* which brings the conductor guide in centre of *can* by which they can be placed in a more neat and regular order making better work, and I believe less waste than when the *cans* are made larger and intended to save labor. Now this sliver is drawn through at a velocity in ratio with draught and speed of front roller which is 153 revolutions of $1\frac{1}{4}$ in. diameter, and the draught is 5.3 and the diameter of back and middle roller one inch.

Then 5.3)153(29 multiplied by $\frac{10}{8}$ equals 36 revolutions of back roller and the draught between middle and back roller is equal to $\frac{1}{10}$th of whole draught, or should be, and if this slubbing roping is intended to go to roving frame, then the hank slubbing for making 20's yarn will be .85 hank or 9.8 grains to one yard with 4.23 turns of spindle

to one revolution of front roller, thus giving 153 multiplied by 4.23 equals 650 revolutions of spindles ; we will
now find out from these what quantity of work can be
done, taking the front roller at 1.25 in diameter, we get
1.25)16(12.8 a guage point for that size of roller, so by
dividing 153 divided by 12.8 equals 12 hanks in 10 hours,
if the machine never stops, we must allow about 25 per
cent. taken off this 4)12(equals 3 equals 9 hanks divided
by .85 hanks equals 10.73 lbs. per spindle and 10.73)175(16
spindles to one delivery of drawing head. Thus showing
how many deliveries and how many slubber spindles is required to produce so many *pounds* per day, of 10 hours.
In adjusting the fluted rollers for this .85 hank the distance between the front and second roller called the *interval* must be set 1.27 inches from centre to centre, and the
top roller will require 20 pounds weight using either solid
or shell rollers and the roping laid spirally around the
bobbin should have 9.1 layers to one inch in length, always
keeping the traverse rail at its minimum speed but
corresponding to the schedule and the hank roping
being made for this. I may here give the rule, sq. rt. of .85
multipled by 100 equals 9.1 layers and should be strictly
obeyed ; there are also many other *rules* by which we get
the answers to these questions abstractly or by cancellation.

I previously stated that this hank roping must weigh 9.8
grains to one yard ; now, to explain this, there are 840
yards in 1 hank, and this measure weighs 7000 grains.
Now what is the weight of one yard? it is 840)7000(8.33
grains, one yard. You will now discover the simplicity
of this rule, when I show you how I get the weight of one
yard of any hank-roping in grains, we have called our
slubbing roping-hank .85)8.33(9.8 grains. Again, it will
be easily seen how to get the draught for the slubber

frame by 9.8)52 grains (5.3 draught, with the rollers.
And the rule for getting the number of turns of spindles
for one revolution of 1¼ roller is sq. rt. of 21 multiplied by
.85 equals 4.23 turns of spindle, thus giving you a precise
method of getting the number of teeth in twist wheel,
draught wheel, lifting wheel and the cone wheel will be
determined by the diameter of bobbin and diameter of
front roller in conjunction with the system of gearing attached, and the rack change wheel is to let off such a
length of move, that will coincide with the pressure given
by the tension, number of layers per inch and thickness of
roping. In making changes, the twist, lift, and rack are
in geometrical ratio, except in the rollers, when the weight
is in direct proportion to the number of teeth in the change
pinion, but inversely to the draught, these being the general features of the machine, and leave this slubbing roping
which is intended for roving frames.

SLUBBING.

We will now see what kind of slubbing roping is required for number 20's, when using an *intermediate frame*
by referring to the schedule it calls for .51 hank slubbing
.51)8.33(16.3 grains to one yard, the drawing being .65
divided by 16.3 equals four of draught, and the front roller running 200 revolutions per minute, with 3.28 turns of
spindle for one revolution of roller giving the spindle a
speed of 200 multiplied by 3.28 equals 656 revolutions per
minute, this being a maximum speed for this coarse roping
however, we will go by this schedule, then the number of
hanks will be 200 divided by 12.8 equals 15.6 hanks, if
the machine, never stops in ten hours, but we will take 40
per cent. off, thus reducing it to 9.24 hanks per ten hours,

a production of .51)10(18.12 pounds for spindle, giving us 11⅓ slubber spindles to one delivery on drawing frame. There seems to be a great increase in the weight turned off from the same number of revolutions on the spindle, still it is no exaggeration when you have proficient help, but the labor is increased with the production requiring some assistance when doffing; to reduce the per cent. of loss, which would incur in these intervals, using all expedient measures to the starting up of the machine that we may derive the greatest benefit from it, for time is money, and if there be a deficiency in single help, and allowing the machine to stand, it is a waste of strength to the help and a loss to the proprietor which will eat like a cancer and assist in making his business unremunerative, for it is always the best to select machines of great productive powers, consistent with the more required merits of doing it well and such as are calculated to relieve the labor and expenses in repairs on the machine. I am in favor of this class of machines because we reduce the quantity, for it is horrid to see this incumbrance of machinery and the production no greater, although it is necessary for the improvement of the work sometimes to use an auxiliary machine like the system we are now considering. We have already seen the benefit of reducing the draughts and the quantity of slubber spindles necessary for the same production by comparison, when not using this intermediate we require 7 per cent. less roving spindles and 80 per cent. more slubber spindles and 50 per cent. more draught altogether. The *drawing sliver* used for this slubber weighs 65 grains to one yard, which is a volume that is most practical to insure a regular and perfect roping. Coinciding with the draught used, which will correspond at all times with the draught capable of being held with the *top* rollers by being kept in closer contact with all its fibres, not allowing any portion of its section in advance of the whole, which often occurs when the sliver is too large, raising the top rollers so high as not to be governed with the edges of sliver exposed, and not retained to be teased out by the *drawing roller*.   It is such little things as these that contribute toward defeating your purpose and detrimental to making a nice level thread. We shall have to increase the weight for top rollers on account of this coarse sliver, to 28 lbs. on front roller, and the same on *back* and *middle* rollers.   The breakage draught here is 1-10th of the whole draught, that is, if the whole be 4, then 1-10th will be 4-10, so that leaves 4 minus .4, equals 3.6, for *front* roller, plus .4 equals 4 draught, although it would be more correct to say 3.6)4.(1.11 equals for back and middle roller as all draughts are as their products then the whole draught will be 3.6 multiplied by 1.1 equals 4, the same. These top rollers are covered with leather, except the back is sometimes an iron-fluted roller, which is not so good, the flutes have a tendency to crimp the fibres and injure them, the hold being too rigid. These leather rollers are often varnished to make them more durable and prevent them from lapping ; this may be allowed on shell rollers, as each *boss* is independent, but where solid rollers are used, varnish with gritty substance in it should be discarded, as it wears off the flutes of steel rollers in a very short time and destroys them altogether, the shell rollers are preferable for the front when having sufficient weight on them, but solid rollers have greater tearing force, and will draw with a less weight, but not so regular.   Now the front roller being 1¼ inches in diameter, will deliver 3.927 inches in one revolution, and the spindle 3.28 revolutions for the above, giving us 3.28 revolutions, divided by 3.927 in. equal .8 of a turn per in. of twist, and by cancellation is sq. rt. of 51 multiplied by 1.13, equal .8 turns the same, the 1.13 being a G. P. for 1¼ in. rollers, this amount of twist being practical for middling cottons, rendering it tensible by the rollers in the next machine. We shall have to make the number of layers for this hank sq. rt. of .51 multiplied by 100, equal 7.15 per inch in length on bobbin. After laying the first layer, the length of bobbin to be made, should decrease one layer every reversion, making a bobbin with conical ends, built in such a manner to prevent the roping from running over, by having a taper of such a bevel, made by this diminution in length of bobbin as it increases in diameter, and this builder wheel should be made changeable according to the number of laps round the *presser*, or the tension of a spring, the greater the tension, and shorter is the move of rack, thus making a short taper and putting more length of roping on the bobbin, this is all right if the ends are smooth and a clean bevel. We have to take this tension off sometimes when the cotton is poor, by taking one lap off the presser, and by doing this the move of rack is increased, and then will make a longer taper and reducing the length of roping on the bobbin, hence a change of the tapering wheel. We must not allow any loss in production through this change, but if the twist requires increasing then it is unavoidable.

No. 5 TABLE

—FOR AN—

INTERMEDIATE FRAME.

Nos.	Hank Sliver.	Grains Per Yard.	Lbs. Per Sp.	Rev. of Roller 10 x 5	No. of Turns.	Layers.	Draught.
7	.64	13	15.75	200	3.67	8.	3.44
8	.68	12.25	14.74	194	3.78	8.25	3.51
10	.75	11.1	13.12	182	3.96	8.7	3.63
11	.78	10.7	12.06	170	4.1	8.85	3.68
12	.81	10.28	11.97	172	4.12	9.	3.71
13	.84	10.	11.5	170	4.2	9.18	3.76
14	.88	9.47	11.	170	4.3	9.4	3.83
15	.9	9.26	10.62	165	4.34	9.5	3.85

Nos.	Hank Sliver.	Grains Per Yard.	Lbs. Per Sp.	Rev. of Roller 8 x 4	No. of Turns.	Layers.	Draught.
16	.92	9.05	10.21	205	4.05	9.6	3.89
17	.95	8.77	9.9	200	4.1	9.78	3.92
18	.98	8.5	9.5	195	4.17	9.9	3.96
19	1.	8.3	9.21	190	4.21	10.	4.
20	1.02	8.17	9.06	188	4.26	10.	4.02
21	1.04	8.	8.66	180	4.3	10.2	4.04
22	1.06	7.86	8.5	180	4.35	10.3	4.07
23	1.09	7.64	8.22	175	4.41	10.45	4.11
24	1.10	7.57	8.	170	4.43	10.68	4.14
25	1.14	7.3	7.8	170	4.5	10.7	4.17
26	1.15	7.2	7.62	167	4.53	10.78	4.18
27	1.16	7.18	7.38	162	4.56	10.78	4.2
28	1.18	7.06	7.18	158	4.58	10.86	4.22
29	1.20	7.	7.07	158.	4.62	10.95	4.24
30	1.22	6.83	6.89	155	4.67	11.	4.27

THE INTERMEDIATE FRAME.

The Intermediate Frame is next in order; its purpose being to get more doublings and reduce the slubbing roping to a finer roving. This machine has a *creel* in which the slubber bobbins are placed in a vertical position, having a skewer through them on which they revolve by their ends resting in a smooth cup and the other ends going in a hole made to keep them upright. These creel rods or plates are made of iron, in which these holes, and cups, or countersinks are made, at such a distance as will allow full bobbins to revolve, but it is best to set half bobbins next to full bobbins, and so on alternately, all through the whole creel. These *slubber ropings* being .51 hank, two of these are put through one guide, which is called *doubling*, and attenuated by the rollers and converted into one roving by the twist, it being reduced from 2).51(.255 hank roping multiplied by the draught viz. .255 multiplied by 4.02, equals 1.02 hank roving for 20's yarn, requiring sq. rt. of 17.8 multiplied by 1.02 equals 4.26 turns of spindle for one revolution of *front roller*, the number of revolutions being 188 multiplied by 4.26 equals 800 revolutions of spindle per minute. The number of layers sq. rt. of 1.02 multiplied by 100, equals 10 layers per inch in length. The quantity of work turned off by this frame will be got from a new gauge point—roller 1⅛ in.)16.(14.2 G. P., a divisor for 1⅛ in. roller then we will get 14.2)188(13.24 hanks per 10 hours without stopping, but for breakages and doffings we shall allow 33 per cent. deduction viz. 13.2 multiplied by .67, equals 9.24 hanks per day. This is like the slubber, a maximum rate and will amount to 1.02)9.24(9.06 lbs. per spindle averaging *two* intermediate spindles to *one* slubber spindle, this being a proper ratio; for this system of preparation which encourages a little more speed, for you are supposed now, to be using a better class of stock, with more doublings warranting an evener roving by the approximation of its fibres, thus giving additional strength to the sectional and longitudinal portions of the thread, by which this excellent frame is so highly recommended. We shall be obliged to close our rollers up a little in this frame, making the interval between front and second, from centre to centre, 1.25 in., also increasing the weight for top rollers to 22 lbs., our experience here being our guide, and guarantee this an average weight from one hank up to two hank roving, you see there are two ends under one boss, and four ends to a roller, this being weighted in the middle by a hook, which connects with the weight-wire and weight. The front top roller should have a diameter as large as the interval will allow, in any of these frames when used as a master drawing roller, giving it more traction surface on the rovings to be drawn and seeming altogether to have a beneficial influence on the fibres, while in the action of drawing, and whenever we see intuitively, these advantages should be mementos in our researches, which will be useful, endeavoring at all times to collect such information, whether it presents itself unlooked for, or acquired by experiment, must be retained for the benefit and progress of human achievements; this machine here under our consideration is a representation of what has been done by the retentive powers and accumulated experience of those who have been persistant in their labors, and unceasing in their efforts to remove that which is unprofitable, by some new invention, whereby we can redeem ourselves by the quantity and quality this invention produces over our primitive methods, each and every improvement contributing towards the perfection of our machines, and reducing our labor.

In reference to these drawing rollers on this machine, it is best to have a moderate draught like the one prescribed, for the task is more difficult on account of the coarse twisted roping, and the number of doublings under one roller, which makes it require such heavy weighting of the top rollers, unless relieved by a greater interstice of front and second rollers, which will be injurious to the roving if the extension exceeds the volume and staple, these two being the connecting links for the measure of intervals, and will be governed in some respects by the amount of twist previously put in these coarse ropings, which requires a little watchfulness by the overseer, for if the roping at times should partially stop the front top roller, and deliver here at a velocity of middle roller, without being attenuated, it is evident then for you to relieve yourself from this dilemma by pursuing the above course ; then again if the coarse roping should be drawn through at a velocity equal to front roller without being attenuated, there is then a deficiency of weight on the back and middle rollers or excessive twist, with too little breakage draught between these two rollers. We are compelled at times to resort to indulgencies, regardless of its issue, when emergencies require immediate relief, so that the machinery may be kept running and prevent disorder in the room by the help. These frames having such heavy weights on the top rollers, endangers the necks of steel rollers by wearing them and the squares out so fast, particularly if the frame is too long by twisting the ends right off, they should not exceed 120 spindles, 8 in. by 4 in. bobbins, this being adequate for a 60 spindle slubber of 10 in. by 5 in. bobbins, when the hank roving-corresponds to the schedule.

In getting the number of twist per inch for $1\frac{1}{8}$ diameter roller is to get number of turns of spindle for front roller, one revolution equals 4.26, divided by 3.53 circumference of roller, equals 1.19 twist per inch, showing that the number of turns are in direct ratio with hank roving, but inversely to number of teeth in the twist wheel, and there should always be sufficient twist put in the roving, to hold and retain it without any loss, until it reaches the rollers of the next machine; sometimes you will discover when taking the flyers off the spindles preparatory to doffing, the roving will not break off as it should do, close to the bobbin on account of too much twist in the roving, and cannot be reduced if the machine has too great a speed on it, unless you bring it to the *limit* where the twist wheel is tantamount to the speed of spindle, which is the proper test for arriving at the greatest production of these machines, and which will maintain its maximum production, by having a system of doffing, like we have for spinning frames, which will reduce the loss in production by stoppages, for slubber and roving frames to twenty minutes for doffing and breakages for every sett of full bobbins made, according to the schedule.

—FOR A—

FINE ROVING FRAME.

Nos.	Hank Sliver.	Grains Per Yard.	Lbs. Per Sp.	Rev. of Roller.	No. of Turns.	Layers.	Draught.
7	1.45	5.74	5.25	157	5.08	12.1	4.52
8	1.58	5.27	4.91	157	5.30	12.56	4.65
10	1.84	4.53	4.375	157	5.72	13.6	4.9
11	1.95	4.27	4.02	150	5.90	13.9	5.
12	2.08	4.	3.99	155	6.1	14.4	5.10
13	2.19	3.8	3.83	155	6.25	14.42	5.19
14	2.35	3.55	3.66	157	6.48	15.32	5.31
15	2.41	3.45	3.54	156	6.56	15.5	5.36
16	2.51	3.32	3.40	154	6.7	15.85	5.43
17	2.62	3.17	3.3	154	6.82	16.2	5.51
18	2.75	3.05	3.16	153	7.	16.6	5.6
19	2.82	2.95	3.07	152	7.1	16.8	5.65
20	2.92	2.85	3.02	155	7.21	17.1	5.71
21	3.02	2.75	2.88	151	7.35	17.4	5.78
22	3.11	2.68	2.83	151	7.45	17.6	5.83
23	3.21	2.58	2.76	152	7.53	17.9	5.89
24	3.3	2.52	2.66	150	7.68	18.2	5.95
25	3.45	2.42	2.6	152	7.85	18.6	6.03
26	3.48	2.40	2.54	150	7.9	18.7	6.05
27	3.55	2.35	2.46	147	7.96	18.85	6.1
28	3.65	2.28	2.39	147	8.08	19.2	6.15
29	3.75	2.22	2.36	148	8.19	19.4	6.21
30	3.83	2.18	2.30	146	8.25	19.6	6.25

THE ROVING FRAME.

The Roving Frame, is the next machine using the inter-mediate bobbins, which are placed in a creel, by which they can be unwound in a systematic manner, by taking two of these rovings through one eyelet of traverse guide, forming one thread, termed doubling; they are then atten-uated to the desired length and weight by this operation, which helps and assists in making a more uniform thread, being an element of the art of cotton spinning, whereby we have produced by previous arrangements and calcula-tions, a *draught* consistent with the hank roving, for it must not be supposed that any draught is right that will bring the doublings to the desired hank, or is sufficient, or that the counts of yarn shall be made from a hank roving that is inconsistent with them, produced either by an ex-tensive draught, or by a reduced one, such methods are to be deprecated. It having struck me years ago, that there ought to be a better way of doing this business, for in one mill they will have two hank roving, making numbers from 8's up to 20's, when I go to another mill, they are using $2\frac{1}{8}$ hank roving, another mill will be using $2\frac{1}{4}$ hank roving, and another $2\frac{1}{2}$ hank roving, and so on. Every one of them claiming to be beating his competitor; I often wonder how such a variety of treat-ments will bring forth such charming results, for every individual manufacturer, under such different ways of obtaining it. Now it is the arranging of the draughts, that is paramount in forming a system by which we can refer to, these call for certain weights and measures to be obtained from correct rules. The 1st being a rule to get the number of hank rovings from the counts to be made, which are No. 20's; here this No. 20 is squared equals

400 divided by 2, equals 200 ; extract the cube root of this, which gives us 5.85 hank roving, this being a roving to be doubled in the spinning for No. 20's, but if these No. 20's are to be made from single roving, then 5.85 divided by 2, equals 2.92 hank roving, which is the specified number in the schedule from intermediate roving.

In putting the rule more precise and arithmetical, it equals cub. rt. of (numbers squared divided by 2), equals hank roving used single.

The rule for draught equals cub. rt. of 64 multiplied by 2.92 equals 5.71 draught, this being the draught required for 2.92 hank roving made from 1.02 hank intermediate roving, and by this rule you can determine any other draught from the hank roving, which you can see are very moderate and will insure large productions by the machines, and all other portions of the schedule being kept up to, that, belongs to this number of counts. We must keep a close watch and adhere to every item prescribed in it, with a desire to learn from it a more methodical course by which we attain superior results.

I shall now give a full description of the roving frame, and how to make the calculations for the different motions, showing how simple it is for the solutions of questions relative to all these motions, in a very concise manner, but, which seem difficult. Now we require for No. 20's yarn a 2.92 hank roving, and the number of turns of spindle will be sq. rt. of 17.8, multiplied by 2.92 equals 7.2 turns of spindle for one revolution of the front roller. And the draught will be 5.71, and the number of layers equals sq. rt. of 2.92, multiplied by 100 equals 16.5 layer per inch on the bobbins, and the ratio of spindle to driving shaft being 3.05, and the ratio of extremes of revolutions of bottom cone equals 3.05. I shall describe from a machine that is most convenient to me, which I believe is one of Higgins & Sons. In timing the front roller I find it running 166 revolutions, this multiplied by 7.2 turns of spindle for one revolution of front roller, equals 166, multiplied by 7.2 equals 1200 revolutions of spindle per minute, a good average rate of speed for a 7 in. by 3½ in. bobbin, and should not exceed this rate unless a much finer roving is made, when the lifting rail becomes reduced in its motions, also the capacity of the bobbins.

The diameter of bobbins, 1 7/16 in. multiplied by 3.1416, equals 4.5 in. circumference, and the roller is 1⅛ diameter, multiplied by 3.1416, equals 3.534 in. in circumference, making the surface velocity of front roller equal 3.534, multiplied by 166, equals· 586 in. per minute, this being divided by the circumference of bobbin, equals 586 in. divided by 4.5 in. equals 130 revolutions of bobbin in excess of spindle, equals 1200, plus 130, equals 1330 revolutions of bobbin when making the first layer, and taking up, 130, multiplied by 4.5, equals 586 in. of roving per minute, which is just equal to the delivery in inches of front roller. Then 1330 divided by 3.05, the ratio of spindle to the driving shaft one revolution, equals 436 revolutions of long sleeve, which I will denote as X at the first layer on bobbin. Now the roller is 1⅛ in. diameter, divided by 3.5 in. diameter of full bobbin, equals .3214, multiplied by 166 revolutions of front roller, equals 53.3 revolutions less when the bobbin is full, so 1200, plus 53.3 equals 1253, divided by 3.05,. equals 407 revolutions of X when bobbin is full. The spindle, 1200 divided by 3.05, gives us the number of revolutions of driving shaft, equals 393 revolutions.

Then X equals 436 minus 393, equals 43 revolutions to be given by the compound motion, this being equal to 43/2, equals

21.5 revolutions of the stud or compound wheel, when commencing on an empty bobbin, its motion being contrary to the driving bevel wheel, when the *bobbin* leads the flyer, and with a slower motion than this driving bevel, "hence" the X will have a speed equal to the sum of twice the revolutions of the compound and speed of driving shaft, equal to 393 plus 43, equals 436 revolutions of X, and by decreasing the speed with the cone, in proportion to the thickness of roving, will give a motion to the bobbin to suit the increasing diameter of the bobbin, and the motions decrease as the bobbins diameter increases.

So when the bobbin is filled the X is 407 minus 393, equals $\frac{14}{2}$ equals 7 revolutions of compound wheel, a speed equal to a bobbin 3½ in. diameter, but when the *flyer* leads the bobbin the compound wheel is reversed in motion only, the same rate of speed is retained, but by its motion being slower than the driving bevel wheel, and rotating the same way, the speed of bobbin is decreased at the first layer on the bobbin, but as the bobbin increases in diameter the speed of X increases at every change until the last layer when bobbin is filled. When you want to change the lead it is only necessary to reverse the motion of this compound wheel, and the flyer presser.

The top cone has a velocity of of 166 multiplied by $\frac{130 \text{ wheel}}{70 \text{ wheel}}$, equal 308.3 revolutions per minute, the diameter of large end of cone equals 6 in. and the small end of cone equals 3½ in. equals $\frac{6}{3.5}$ equals 1.72 ratio multiplied by 308.3 equals 530 revolutions of the bottom cone when driven by the large end, and on the contrary, will be 530 divided by 3.05 equals 174 revolutions of the bottom cone. Then 530 minus 174 equals 356 revolutions for 3½ in. bobbin or $\frac{356}{540}$ multiplied by 30 in. the length of cone,

equals 20 in. the belt as moved for a 3½ in. bobbin, although this could be changed by using a different cone
wheel, and would alter the travel of cone belt. When
once the cone wheel is ascertained so that the tension of
roving from roller to flyer is right, should never be
changed at any future time on account of different hank
rovings, for it is wrong and unnecessary. Now $\frac{3.5}{2}$ equals 1.75
in. minus $1\frac{7}{16}$ divided by 2, equals .718, equals 1.03125 in. the
depth of cotton on side section of the bobbin, the num-
ber of layers per inch being 16.5 multiplied by 1.03125
in. equals 17 layers in depth, but, the tension and pressure
call for 4 times that multiplied by 17 equals 68 reversions for
twice round the presser and 4.5 times that multiplied by 17
equals 76.5 reversions for three times round the presser, however, we will say twice round, then the number of revolutions
at every change becomes 356 divided by 68 equals 5.2
revolutions less, and 20 in. divided by 68 equals .29 in.
move of rack, and 1.00 in. divided by .29 equals 3.44
moves in one inch in length of cone, and 3.44 multiplied
by 5.2 equals 17.8 revolutions less per inch of travel giving us 356 divided by 17.8 equals 20 in. travel of the cone
belt, and 16.5 layers multiplied by 4.5 equals 74.25 in. to one
in. in length on the empty bobbin and 7 times that equals
520 in. length on.   If 586 divided 74.25 equals 7.9 in. length
per minute laid, then 7.9 in. divided by 8¼ in. circumference of lift wheel equals 9.57, revolutions per minute
and 7 in. lift divided by 8¼ in. equal .848 revolutions of lifting shaft.   The number of extra revolutions
of X 130 minus 53.3 when full equals 76.7 revolution less,
equals one revolution for every reversion three times
round the presser, and by 3.5 multiplied by 3.1416 equals
11 in. multiplied by 16.5 equals 181.5 in. per inch in
length, this multiplied by 4.12 in length. of last layer

equals 748 in. Then 586 in. divided by 181.5 in. equals 3.228 in. per minute in. length and 68 layers divided by 16.5 equals 4.12 in. length of lift on last row. If the cone belt as 20 in. travel then $\frac{20}{29}$ equals 68 reversions, and the length of extremes multiplied by the half of reversions will equal length of roving on the whole bobbin in yards. The first layer 520 in. on 7 in. in length, the last layer being 748 in. on 4.12 in. in length, the sum of extremes being 1268 multiplied by $\frac{68}{2}$ equals 34 multiplied by 1268 equals 43112 divided by 36 equals 1200 yards of roving on bobbin. But if the roving be wrapped three times round the presser then the length will be equal $\frac{1268}{2}.\frac{76.5}{36}$ divided by 840 equals 1.6 hanks on the bobbin 7 in. by 3½ in. And now I think we have got it pretty well understood so that we know what and when a change ought to be made. If the cone wheel is changed the rule is cone 2 : cone 2 :: Rack wheel required, for if the cone wheel requires to be larger it will draw the ends taut, and so will a larger rack wheel, but the cone and the rack wheels are inversely to each other, that is if I put a larger cone wheel on, I require a less rack wheel, and one tooth at the cone will equal one lap round the presser directly, and one tooth of change at the lift-wheel, will cause a change of two teeth of rack being in direct proportion to the lift-wheel. So, by putting a small cone-wheel on, we get more reversions, consequently, more cotton on the bobbin, and the more laps round the presser means more cotton on the bobbin, and the lifting-wheel should never be changed on the same hank-roving, which is often done. I will now explain, by the wheels on this machine, how we get the number of teeth in the cone-wheel, for a bobbin $1\frac{7}{16}$ in. diameter, for it is through the change of diameter of bobbin alone, which infers a change of cone-wheel. We will start at the compound-wheel, which has 110 teeth,

requiring 21.5 revolutions per minute, given by a pinion
gearing into it of 14 teeth on this same shaft, is a 50 gearing into a 60, connected with this, is an 80 wheel gearing
into cone-wheel, this bottom cone has a velocity of 530
revolutions.   I will now get the ratio of these wheels
from cone wheel to compound wheel, which are seldom
changed, $\frac{80.}{60.}\frac{50}{14}$ equals 4.76 ratio.   Now then, $\frac{110\times21.5\times4.76}{530}$,
equals 21 cone-wheel required for a bobbin $1\frac{7}{16}$ in. diameter.
The lifting motion or change wheel required for this machine, is by taking a cancellation of the train of wheels
from cone to lift shaft, and is simply 530 divided by 16.5
equals 32, wheel on the end of long shaft going under
frame.   Now this 32 change wheel for 2.92 hank roving
is calculated from the use of the 35 and 76 teeth bevels in
this train of gears.   But these bevels are sometimes changed
to 44 and 54 teeth bevels, which will alter the change
wheel, by dividing change wheel, 32 divided by 1.76 ratio
equals 18 wheel required on the end of long shaft, this
being a ready mode of getting the change wheel on Higgins & Son's frame.

I do not think it is necessary to go into a full detailed
description of this lifting motion as every machine maker
as a different train of gears from cone to lift shaft, and it
would be superfluous in hunting up such an history of
changes, and is not required by the overseers or manufacturers, these calculations having been made by the machine maker, before he sends the machines to the factory
to be operated on, however, I have given you the rule how
to get the number of layers per inch for any hank roving
you wish to make, and from a trial on the machine which
will show you how many layers per inch it is geared up
for, this being compared with the one required, you can
easily determine by proportion what change wheel is,

necessary, by actual demonstration. I have yet to ascertain what twist wheel is required to give us 7.2 turns of spindle for one revolution of front roller. There is a wheel on the end of front roller of 130 teeth which gears into another of 70 teeth on the end of top cone shaft, on this same shaft, there is a wheel of 40 teeth that is geared with the twist wheel by a carrier, and the ratio of spindle with driving shaft is 3.05. So, then, by a little canceling, we can obtain the number of teeth in the twist wheel required from the data or wheels given $\frac{130}{70} \cdot \frac{40}{7.2} \cdot \frac{3.05}{1}$. equal 31 twist wheel.

We have now to take the draught of the rollers from the wheels and diameter of fluted rollers generally used on the roving frame. We will commence at the front roller, which is $1\frac{1}{8}$ in. in diameter, and the back roller 1 in. in diameter, having a back roller wheel of 56 teeth, which gears in the change pinion. On this shaft is a stud wheel of 100 teeth, gearing into a 30 front roller wheel, and from these is $\frac{9}{8} \cdot \frac{56}{5.71} \cdot \frac{100}{30}$ equals nearly 37 pinion. The schedule gives the draught 5.71 requiring a 36 or 37 change wheel and is obtained very readily by canceling, thus: G. point 210 divided by change wheel equals draught. This method can only be used on this arrangement. But you can easily make gauge points for any system of gears used on other machines by the same mode of procedure. We get the ratio of spindle from the following wheels necessary to that result. On the driving shaft is a 42 teeth wheel which drives a 28 teeth on the end of spindle shafts, on these are wheels of 55 teeth driving 27 teeth on spindle, so $\frac{42}{28} \cdot \frac{55}{27}$ equals 3.05 turns of spindle for the driving shaft one, and by turning this shaft round until you have made one revolution of front roller, you will discover that the spindle has made 7.2 revolutions in the same movement, this being in ratio with a 31 twist wheel.   I may also state that this wheel is coincident with the lifting wheel for any hank roving you wish to make.

N. B.—These calculations are made from a frame running 1200 revolutions of spindle.   You must go by the schedule and not from this experiment.

No. 7 TABLE,

—FOR A—

COARSE ROVING FRAME.

Nos.	Hank Sliver.	Grains Per Yard.	Lbs. Per Sp.	Rev. of Roller.	No. of Turns.	Layers.	Draught.
4	.93	8.9	6.76	152	4.07	9.6	4.46
6	1.22	6.8	5.99	162	4.67	11.	4.89
7	1.34	6.2	5.58	161	4.9	11.55	5.
8	1.46	5.7	5.26	160	5.1	12.1	5.19
10	1.7	4.9	4.66	157	5.5	13.	5.4
11	1.8	4.65	4.3	150	5.67	13.4	5.56
12	1.93	4.36	4.29	160	5.85	13.9	5.7
13	2.03	4.1	4.1	158	6.	14.2	5.79
14	2.18	3.85	3.926	162	6.25	14.8	5.93
15	2.23	3.74	3.79	157	6.30	14.9	5.98
16	2.32	3.6	3.655	155	6.45	15.2	6.05
17	2.42	3.45	3.5	153	6.51	15.5	6.14
18	2.55	3.3	4.40	157	6.75	16.	6.24
19	2.62	3.2	3.286	153	6.85	16.2	6.3
20	2.7	3.1	3.22	155	6.95	16.45	6.37
21	2.8	3.	3.09	152	7.08	16.72	6.44
22	2.9	2.9	3.02	154	7.2	17.	6.52
23	3.	2.8	2.935	153	7.3	17.3	6.6
24	3.1	2.7	2.85	154	7.45	17.6	6.67
25	3.2	2.62	2.783	153	7.58	17.9	6.74
26	3.22	2.6	2.71	150	7.6	17.95	6.76
27	3.3	2.54	2.635	148	7.7	18.15	6.81
28	3.36	2.48	2.568	147	7.8	18.32	6.85
29	3.45	2.42	2.52	147	7.95	18.6	6.91
30	3.55	2.35	2.45	140	8.	18.85	6.97

I will return now to the first system of making No. 20's
yarn, by dispensing with an intermediate frame, and using
the .85 hank slubbing roping on the roving frame, according to schedule which calls for a 2.7 hank roving, being 7
per cent. coarser than the other method. The draught
being cu. rt. of 96 multiplied by 2.7 equals 6.37 of a draught,
requiring 210 divided by 6.37 equals 32 or 33 change pinion, reducing the weight from 9.8 grains to 3.1 grains per
yard, with sq. rt. of 17.8 multiplied by 2.7 equals 6.95 turns
of twist for one revolution of the front roller, and sq. rt. of
2.7 multiplied by 100 equals 16.4 layer per inch in length
on bobbin, and twist wheel of 33 teeth and a lifting wheel
the same, 33 teeth, and a cone wheel of 21 teeth, with a
move of rack of .27 in. every change, which can be let
off by using a 11 star wheel with a 30 fastened on it, and
a change rack wheel of 48 teeth. The weights for the top
rollers where there are four ends under one roller ought to
have 18 and 20 pounds, and is sufficient up to four hank
roving, and the interval between front and second rollers
from centre to centre, should be 1.20 in., and the front roller
having a velocity of 155 revolutions per minute, with 6.95
turns of twist for one revolution of front roller, making
the spindle have 155 multiplied by 6.95, equals 1000 revolutions per minute, and producing 14.2)155(10.7 minus 17
per cent. equals 8.69 hanks for ten hours, and a weight of
2.7)8.69(3.22 lbs. per spindle, and 6.95 divided by 3.53,
equals nearly two turns per inch, and the number of roving
spindles required to one slubber spindle are 3.3, or the
number of slubbing spindles can be determined upon by
multiplying number of roving spindles by .3 equals the
number of slubbing spindles.

I think we have got the rules for making the changes

when required, pretty well understood now, and I hope they have left an impression on your mind that will always induce you to act in no other way than those previously given, showing that you have a system about your business that makes it a science with you, by putting in practice every rule and law given in schedule, also with the knowledge of things which are here propounded and advised.

It would also be a fit time now to say a word or two about the flyers before using them, as they have such an influence over the production of the frame, when running at such high speeds, by having the pressers put in proper shape, and should be kept so, which must be similar in shape, every one of them to be successful. Now I propose blocking every presser on the flyer before being put in use, this is done by taking the flyer in the left hand and placing it in a block resting on the overseers bench ; this block is made in such a form, as to put the presser in a proper shape, and it also binds it on the flyer leg at the same time, by giving the top block a blow with the hammer as you are holding it in place with your left hand. These pressers are easily put out of shape, and it becomes necessary to have these blocks at the bench, so that when one of them gets bent out of shape, it can easily and readily be fixed, even by the minder herself. The importance of this will be discovered by one who understands the cause of slack and taut roving. For when the drawing and slubbing, and roving are all even, there must be some other cause which will be found in the presser being out of shape. This block puts them all in one shape, causing them to press the roving so regular, and making every bobbin have the same diameter, " hence good roving is the result, besides increasing the production by preventing stoppages by broken ends, and changing wheels all the time; in fact they are indispensible when quality and quantity is considered, for any irregularity in such delicate work as making roving is destructive to all the previous manipulations of it, and will certainly be the cause of making weak and uneven roving."

The principle of this block is to get all the pressers so that they will come into contact with the bobbin at a point on the presser, which will always get the same length of leverage from the flyer, producing an even pressure, and causing the tension of the roving between the bobbin and the roller on every end alike, and when once regulated it is done with, giving the minder less work, and making less waste, also saving the overseer the trouble of so much changing. These pressers are apt to get bent and broken, and replaced, by new ones being put on, and if we had not this block we should soon have them in all kinds of shape, for the eye cannot detect this small difference, and are often let go for want of evidence in not being able to distinguish such a small difference in their shape, and it is from that very thing, when you call the work running bad which produces it. There is also another point about this blocking, which prevents the pressers from getting all bruised and hammered up, for they leave this block without mark or blemish, securing a smooth roving, for when in the block it is put in the proper circle by a blow on the top block, which gives it by percussion the required shape, as it cannot be easily accomplished by any other pressure, for it acts like stretching by peening, and if they are done by pressing or squeezing it springs back again, and so far as my experience has been with all kinds of pressers, I would certainly prefer a single centrifugal to any other extant, by its being simple and requiring less repairs than double or spring pressers, although the latter will press more cotton on the bobbin, yet you cannot make two springs have the same pressure in this case, hence more twist or irregular roving is the result. The single centrifugal are much better to handle by the minder, but they should be properly hung to the flyer, and of equal weight, being easily adjusted and replaced when one breaks, and should be so applied to the flyer, that when not rotating the finger will fall to the bobbin, and when in motion it should have no desire to fly back on account of the leg of presser not being in its proper position, for so soon as it receives motion, the weight of the leg of presser should obey the law of central force, by its flying out tangentially and its force increasing as its rotary motion increases in an opposite direction, by the finger of the presser being at right angle with the leg, and its centripetal force will be equal to momentum and the leverage of the finger ; see that they all hang loose and not to bind on the flyer round the bottom of the leg, or where it is hung. These flyers ought to be well dished out at the top, at such an angle that will give plenty of friction on the roving, and cause the twist to run up to the front roller, and the slot down the flyer leg should be twisted in order to prevent the roving from running into the slot, and all the time choking it up, for its tendency is to fly to the farthest point inside the flyer leg when running. We will now put the flyers on the spindles, taking care to have the front row reversed to the back row, as the single centrifugal presser get a little out of balance, as the finger is pushed out by the increased diameter of bobbin, and this should be strictly attended to, to secure stability of the machine, for if neglected, the great speed at which they are run increases the danger, and ultimately loosens all the joints, causing wear and tear unnecessarily, and ruining the frame entirely ; this should be strictly attended to, for the minder is subject to doing as she has been taught, and that very indifferently with some, but any way this thing should not be allowed, for they are not aware of the consequences. Let the roller weights be hung with the 18 pounds on the front rollers, and the 20 pounds for the middle and back rollers, it is not compulsory to have the under clearers on, when the roving gets finer than two hank roving, but wherever they are used, do not forget to oil the lace where it rests on the front roller, for it creates great friction on the front roller if neglected. These twenty pound weights are sufficient for the volume of this sliver, and so is the interval, which must not be altered on any account when using middling cotton, for this same volume or hank roving, which governs it altogether when using the same grade or staple of cottons. We will now see that the cone belt be properly joined by being cemented and keeping the surface perfectly even, which are reverse when laces, belt hooks, etc., are used, for this driving is very important and should run evenly on the top and bottom cones, without any jumping or pressing too hard against the belt shipper—this being in such a position as not to let the belt move laterally when in motion, and kept at equal distances and in line with the ends of cones. These cones are 30 in. long but we only get about 27 in. of moving surface when utilizing the whole of the cone, we have proved in our calculations that a 7 in. by 3½ in. bobbin only requires the cone belt to move 20 in. on the surface of cones, which would be much better if the cones were longer, but that would cause a change in the concave

and convexity, or hyperbolic curve of these cones. I have
already seen cones 36 in. long, utilizing 34 in. of the surface, making 7 in. by 3½ in. bobbins. Suppose now we
had 34 in. divided by 68 reversions equals .5 in. move,
instead of .29 in. this is a great advantage where such precision is required in the motion of the bobbin. The driving
of these machines should have such size of pulleys as is
most conformable with the power required, as the machine
is apt to have an irregular motion when there is not a balance wheel large enough to prevent it, and this may have
the balance of power, but the loss in time is against it,
when there are sudden stops and starts to be made to facilitate the piecing up of broken ends, and also in bringing the flyers to that particular position where the girl can
get her hands in best between the flyers and bobbins.
Now in place of large balance wheel I would prefer larger
pulleys, so that the belt will have the regulating power as
well as the motive power, by this change there will be less
accidents to the different motions of the machines, and
will also save a little of the expense of belting by requiring less in width. If we take ⅓ off and make the pulleys
⅓ larger in diameter, we have just the same power. But
it is the governing power that I refer to, more than the
driving, for by its increased radius and traction there is
less chance of slipping or twitching in its motive power,
and enhancing its regulating power, whereby we arrive at a principle which should be consistent with
the nature of the machine and the work it has to
perform, and if this suggestion should increase the cost,
it will be the least when compared with the dilatory and
sluggish action of the smaller pulleys, in not answering to
the wish of the operator when the belt will squeak and

jerk to the annoyance of all around. I have often wished
for this unpleasant thing to be remedied, before coming to the mill, for it is no recommendation to have the
change made there. Now, with regard to the power this
machine requires, is a great consideration and should be
brought to its minimum, by all and every means that can
be applied or reduced, the steel rollers, and top rollers
when weighted, seem to absorb a good deal of it, and it
is a question with regard to the bearings in the stands
which kind will reduce the power of those that are squared
out, or those that are circled out, I should prefer the last
named, because they will hold more lubricating material
on their journals, resisting the friction caused by the
heavy weights on the top rollers. But, which take the
least power, I have not tested; nor the spindle shafts
which run under centre of foot step, which I think
is in a better position by dispensing with skew bevels altogether for driving the spindles. The hubs of the coup-
lings bevels would be improved very much if they had a
saw gate through one side, and then a collar slipped on
with set screws in, would hold and keep the shafts true,
dispensing with the other set screws which are troublesome
and in the way, on top rail shaft when the guage
is narrow. I must not forget the value of having the
conductor rod as close as possible, and the eyes put
pretty close together so that when the traverse is at its extremes, it will let the outside eye overlap the inside one
and keep the surface of the leather smooth and last longer,
a feature to be courted. Now then if the overseer wishes
to put a little of this theory into practice and find out
whether this advice has come from a judicious and careful observer, as he argues to be, and whose experience
has been varied and brought him to such conclusions, from

test and long trials that he feels ready to present them to
the public, to be criticised by the (great schoolmaster) and
when coming from such a desire, it behooves us to try and
prove whether his figures and rules are consistent with the
kind of yarns you intend to manufacture, that is low and
middling counts to suit the general trade around us. It
is expected from you in your experiments to adhere to the
schedule in every branch of manipulation so that its tenets
may be fully carried out, and measured by its results. It
will be as well here to have something to say about the
twist, which is one of the two things most essential in
the science of cotton spinning equal *twisting* and *drawing*.
For the amount of twist required is economy, but when
put in injudiciously it is extravagance, and will even-
tually prove disastrous, where sharp competition meets
us, then it behooves us to be diligent and wise in
applying it both in just what is necessary, to give us both
in length and strength its greatest quantity. The rule for
twist (depends upon the length of fibre and the sectional
area of the thread inversely), and with this motive in view,
we will commence right here at the slubber, which is the
first machine to give twist, there being 4.23 turns for one
revolution of 1¼ in. diameter, front roller, and 3.927 in.,
circumference, which gives 1.04 in., twist per inch when
divided into 4.23 turns of twist, for .85 hank slubbing, and
the grade of cotton previously mentioned, this twist which
is sufficient when your drawing has been properly prepared,
and let there be no changing of this quantity, under any
plea whatever, for excuses are ruinous, and are made
chiefly when there has been some previous tinkering and
tampering done, without any foresight as to the results.
Carelessness and incompetence in this branch of the business, will bring in its course, all those complaints which

succeed in its onward progress to enormous proportions, by the labor and machinery it has to pass through, placing you in a deplorable position, by not being able to present in the market, an article that will command a remunerate price for your labor, and above all, the climax will be, that you have been unsuccessful. Now the rule given for this quantity of twist must be adhered to, for all roping made from $1\frac{1}{4}$ in., diameter of roller, but, when made from $1\frac{1}{8}$ in., diameter roller, it is 6.95 divided by 3.53 equal nearly two twists per inch, plus the latent twist in the slubber roping equal $\frac{1.04}{6.37}$ equal .168 plus 2 equal. 2.168 the whole twist in 2.7 hank roving, then the draught for No. 20's equal 7.9, so 7.9)2.168(.274 plus 20.8 equal 21.-074 twist per inch, for No. 20's ring yarn. I hope these calculations will prove correct, when experimented on, for I don't believe in guess work, for a small fraction of increase twist, or one spindle stopped, will tend to increase the cost of production, and it is these things that seems small but they accumulate to large proportions in the long run, and if you are not expert you will be some time in finding it out what the reason is, and where these discrepancies are, for you think you are on an equal footing with your competitor, with regard to raw material and machinery, and why should you not make as good an article and as much of it in the same time? I say you can by strict attention to these small things, beat your neighbour if he neglects them, so you will now perceive the object of putting just the amount of twist, and how to get at it in a thorough, and mathematical manner, so that there can be no error made at this point. .The weighting of these machines will be found in the schedule, and are already mentioned particularly, they are not to be excessive, because they require more power to drive them, but still there must be sufficient to hold the cotton, and not allow it to be pulled through in chunks, neither should that object be relieved by greater intervals which would entirely spoil it, sometimes, it comes from the rollers undrawn, by being overtwisted, and the girl should always piece her end up from the bobbin, and if the weight is not heavy enough when the *interval twist* and *draught* are put to the schedule it will be the best thing to do to make them heavier but not until you are fairly convinced that the above mentioned are right. You will observe that there may have been some mistake in the assortment of the cotton, which will prevent the schedule weight answering, and if this should be the cause, it will not be necessary to increase the weights, but resort to the next best thing, by relieving the weight on the middle roller, if it be twisted roving, and give it to the back roller, just enough to get out of the dilemma, but be sure to move the hanger back again, when the trouble is over. The slubber and roving frames must have dead weights, and should have back and front alike to prevent mistakes in hanging them on, and it should be distinctly understood, when ordering the machines, that the numbers intended to be spun, should be specified, as the weights are governed by the Nos. in the schedule, and the hooks and saddles ought to be made to cause the least friction, no matter how it is prevented, whether they be wood, or iron, or brass, or any other metal, the one that is best is the cheapest at cost. I would prefer having the weights made after trying the machine, if the material be out of the general line, such as dyed cottons or merinos, etc., which requires them heavier than gray cotton does, and it would be cheaper to make them to order. The balance weights for the top rail should be sufficient when the bobbins are half full, (exclusive of friction), causing them to act favorably with the cone belt; and the weight for moving the driving belt, by shipper rod when necessary, should be ample, and act spontaneously, should there happen to be a mischange with the reversing shaft, and respond at once to the dogs, that unlatches the lever. Hoping that the importance attached to this weighting the rollers may be properly understood and made reliable, to secure perfect tenuity at all times, without having any suspicions that the weighting is at fault, should there come any kind of a snag which appears to show imperfect drawing, let there be no surmise or doubt, "for the weighting depends on quality of *fibre* and *volume* of sliver and interval." The draughts "depends on the parallelism of the fibres, with the volume of the sliver in conjunction with length and fineness of fibre and yarn." These are the consisting elements as to how far the extension of elongation, and attenuation may be carried under these controling elements, which are according to a principle laid down, that demonstrates by actual experiments a result, approximating to perfect drawing.

There are exceptions to be made from this general rule where the numbers often varies on the same machines, however, they ought to be strickly adhered to as close as possible, for this wavering and rambling away from established principles should be denounced whenever detected by those who are competent and have the power to discharge these duties, this being another of the small things I have referred to before, and causes trouble and worriment when your yarn proves bad and difficult to discover, where there is a large quantity of machinery, for this changing is usually done to remedy some previous neglect of the stipulated weighings. We must commence with small draughts, while the sliver is large and its fibres tortuous, but as they become attenuated by the excess of draughts over the doublings, they will also become closer and more parallel as the material advances, so shall the draught be increased through its whole progress, and while arranging these draughts, it is important to know that the pitch of the wheels used for drawing cotton with fluted rollers, should be as fine as the machine and work will permit of, and where there is carriers to be used, always place them over the rollers, and have them conveniently large, as small ones are injurious and inclined to twitter the rollers, a very serious action, and must be guarded against by adopting the above device, or any other that you may suggest, that will surely prevent the evil, and make a radical cure of it entirely. I believe when the roller stands are made to incline towards the front, will prevent the rolling in the necks or journals, and should be used on such machines that are not heavily weighted, corresponding with the fine teeth recommended for the draught wheels, which are limited by the power required for their strength. And the products of these draughts divided into the product of the doubling, will give a quotient equal to the counts or weights. The doubling " depends on the quality of fibre and directly to the length and strength, and (*vice versa*) to its shortness and weakness." It is intended to improve the unevenness of the sliver or roving, by getting an equal number of fibres in section of thread through its whole length, which is supposed to be called a level thread, and giving it equal strength in every portion, by having got as many fibres equally distributed as possible in the thickness of it, being also better prepared to receive increased draughts, when the cotton is of such a quality to warrant it. But sometimes we get inferior grade of cotton that will not stand

the doublings, consequently we cannot draw so much,·
"hence coarse counts." And if not very even, the strength
is increased by the tortuous fibres being so linked with each
other, that it modifies or helps a weak and short stapled
cotton to a certain extent; but if we attempted to
make a fine thread from a twisted roving, whose
filaments were crossed, we should surely fail, and
especially if we endeavored to do it by drawing in the
rollers, it can be done by being partially drawn and
stretched in the spinning mule, but not alarmingly, although it will help to even it and make it more salable.
It is in this machine (mule) only, that such kinds of cotton can best be spun, on account of it having the two
functions at the same time, viz.: drawing and stretching,
being free from any drags, attenuating and twisting it to
the full extent of which its fibres will allow. Claiming
this advantage over the spinning frame, of remedying by
stretching what the latter needs, by doublings, also enabling it to work up a large portion of waste, which ought
to be used sparingly when mixed with raw cotton to be
spun on the ring frame, for it will surely show itself in
puffs and spongy places, and the drawing utterly failing to
accomplish the task of making the thread even by any of
the machines it has to pass through.   Now we will try to
show how such stock, composed of such unequal lengths
and diameters in their fibres, can be drawn.

I mentioned, when treating on drawing cotton, that
only two rollers were necessary to make a whole draught,
but in this compound of material we are obliged to use
auxilliary rollers and weights, at intervals and pressure in
a ratio suitable to the unequalled fibres and proclivities of
their inherent nature, of which we must partially understand before we attempt to adjust these several require-

ments in excess of pure cotton drawing, or we attempt to do something more than we know, exhibiting a good sample of ignorance in the experiments and producing nothing but a waste of time. I say this from my own experience, wishing that others may benefit from it and save them that labor and worriment which often enfeebles that energy of mind required, in emulating and trying to excel in this, over your competitor, by some novel or expeditious way of making an article equally as good at a lower price. While such is consistent with our ingenuity, yet it is even as deplorable to see the monomaniacs and gray-haired equally as prolific, by the assiduous labor and intense desire to make money faster than your neighbor and to there grave in double quick time, instead of keeping cool and acquiring knowledge, which tends more to your happiness than all your riches. In showing how this kind of material can be drawn, I will illustrate my method with some merino mixings, the short fibres we will call 1.25 in. long, and the longest 2 in. for example, we should by these have a geometrical mean of 1.6 in. this would be the interval between front and second rollers, I would then make the interval for third and fourth rollers 1.7 in., supposing this to be on a double draught drawing frame, on the first head, using a saddle that I can move the weight hook to and fro, as it seems best to draw, giving the feeding roller a chance to let them slip through by easing the weight off a little and moving the hook towards the drawing roller, just enough to obtain good drawing. We will call the whole draught four, making sq. rt. of 4 equal 2 the geometrical mean for each draught, that is two between front and second, and two between third and fourth rollers, and the second and third rollers having equal motions for between these two rollers there is a wide interval, so that the web can be collected and condensed in order that it will hold its fibres together better when passing through again, between the front and second rollers. Now in the second head I would have a double roller stand with three lines of rollers in the front section, so that you can approach the drawing roller, by having the front roller weighted separately, and the second and third by a saddle, that the weight hook can be moved at will, placing the weight where it is most useful, so that the long fibres will escape being torn, and the whole draught remaining the same, two in the back and two in front sections of rollers, and using no coilers at all. It is now ready for slubber, with three lines of rollers, being weighted in the ordinary way, only a little extra weight on, and the second and third rollers with a saddle that the weight hook can be moved to suit the material, and the flutes of the rollers to have 50 flutes to $1\frac{1}{8}$ in. diameter, and to be cut irregular. The front and second roller can be set $\frac{1}{16}$ in. closer in the interval, and relieving the weight on each progressive machine, making the front and second rollers on each a little closer all the way through, keeping the draughts as low as possible, the velocity of drawing rollers being about 30 feet per minute, and the slubber spindles should be 50 per cent. slower than when using all cotton, and the other frames about 30 per cent. slower, the intention is to keep the long fibres from flying out when drawing and putting the twist in. I will, at some future time give a full description of this kind of work, it being much different in its treatment, and will require a full treatise on the subject. I only adverted to it for the sake of illustration in working these mixed materials.

No. 0 TABLE—FOR A SPINNING FRAME.

Nos.	Grains per Hk.	Lbs. Per Sp.	Rev. of Roller	Fine Roving Draught	Coarse Roving Draught	Ring Twist	Mule Twist	Reeled Yarn Twist	Weft Twist	No. of Sp. for 1 rov. F. R.	No. of Sp. for 1 rov. C. R.
4	1760	8.	116			9.3	7.62	7.	6.5	4.75	5.07
6	1132	5.8	114			11.36	9.35	8.6	7.97	5.8	6.2
7	1000	5.0	113	5.26	5.62	12.3	10.	9.28	8.6	6.3	6.7
8	880	4.4	112	5.76	6.15	13.12	10.75	9.9	9.2	6.7	7.18
10	700	3.5	110	5.9	6.3	14.7	12.	11.1	10.21	7.5	8.02
11	638	3.07	109	6.1	6.52	15.4	12.56	11.6	10.75	7.86	8.41
12	582	2.93	108	6.28	6.7	16.1	13.2	12.15	11.21	8.2	8.8
13	540	2.68	107	6.4	6.85	16.72	13.65	12.6	11.7	8.6	9.18
14	500	2.48	106	6.5	6.95	17.4	14.25	13.1	12.12	8.9	9.5
15	467	2.31	105	6.78	7.2	18.	14.75	13.5	12.5	9.2	9.85
16	438	2.15	104	6.9	7.4	18.5	15.2	14.	12.95	9.5	10.2
17	411	2.02	103	7.	7.5	19.15	15.7	14.4	13.4	9.8	10.42
18	390	1.9	102	7.14	7.62	19.6	16.15	14.86	13.75	10.	10.75
19	368	1.79	101	7.3	7.8	20.1	16.6	15.25	14.15	10.3	11.02
20	350	1.71	100	7.4	7.9	20.8	17.	15.65	14.5	10.6	11.3
21	333	1.60	99	7.5	8.1	21.2	17.45	16.	14.85	10.83	11.6
22	317	1.53	98	7.7	8.2	21.85	17.76	16.4	15.2	11.15	11.88
23	304	1.45	97	7.8	8.35	22.2	18.25	16.75	15.5	11.35	12.15
24	292	1.38	96	7.97	8.5	22.63	18.65	17.15	15.9	11.6	12.4
25	280	1.32	95	8.	8.53	23.2	19.	17.5	16.2	11.82	12.65
26	268	1.26	94	8.15	8.7	23.6	19.4	17.75	16.5	12.1	12.9
27	258	1.20	93	8.25	8.8	24.15	19.8	18.2	16.85	12.3	13.18
28	250	1.15	92	8.35	8.9	24.6	20.08	18.5	17.15	12.5	13.4
29	242	1.11	91	8.42	9.	25.	20.25	18.85	17.5	12.75	13.63
30	233	1.06	90	8.5	9.1	25.4	20.82	19.2	17.75	13.	13.88

We have now arrived at the spinning machines, which converts the roving into yarn of any or every description required, according to the preparation and the kind of machine to spin it on. Our intention is to make No. 20's yarn, either from fine roving, equals 2.92 hank, or from coarse roving, 2.7 hank, which can be done by a small difference in the draughts of each, equals 7.9 for coarse roving and 7.4 for fine roving, these draughts corresponding with the different hank roving. The twist in the yarn of both being the same, equals 20.8 twists per inch, and weighing .416 grains per yard, or 350 grains to one hank or 840 yards, requiring 70.14 pounds to break one lea, made on a 54 in. wrap-reel. The speed of front roller to be 100 revolutions per minute, and the twist 20.8 multiplied by 3.1416, multiplied by 100, equals 6540 revolutions of spindle averaging 1.7 pounds per spindle per week of 60 hours, on the ring frame, but, if spun on the mule, the production will be 20 per cent. less, equals 1.37 pounds, and the proportion of ring spindles required for one roving spindle, equals 11.3 for 2.7 hank roving and 10.6 for 2.92 hank roving. These speeds will prove to be, by a very little experience, about as good an average for quality and production that you can determine on, for realizing the most profit. When everything is taken in consideration, it is not very high speeds with smaller spindles altogether, that gives you the best results, it is a coalescence of every respective functional power which this machine will give from its construction, every motion claiming its proper ratio, from the main one, whose merits relatively combined, which must be obtained, in a certain measure, from the material and counts to be spun. I mean the centralization of its greatest powers, whereby we obtain the above mentioned speeds, the results of great experience from all par-

ties who h- /e investigated this matter thoroughly, hoping
that these remarks will prove of some service to you, by
saving you the trouble and expense of having to repeat
them, clinging to the precepts of the schedule from the
start, your future will be a success. That is my object
which I never lose sight of, when trying to describe to you
the most direct route by which you can arrive at success,
keeping in the path I have advised you, for if you begin
to wander, either right or left, from the straight and given
course, you will surely get lost, and get in such a labyrinth
of ways that you will find it very difficult to extricate yourself from, leaving you an unsuccessful man, by deviating
from the course you have been advised to. In setting the
rollers for these ring frames, the interval should be one in.
from front to second, that is from centre to centre, and
the thread guide should be from 2 in. to $2\frac{3}{4}$ in. to top
of bobbin, this being generally fixed by the machine makers;
and it is necessary, when adjusting the frames, always
to set the rings concentric with spindle, and in doing so it
requires the spindle rail to be moved in some cases, when the
rail lifters have worn, and caused it to get out of centre.
Having got these right, then the guide wires can be set
with the spindle, taking care that the balance weights just
exceed the rail, but when you make a doubling frame of
it, the balance weights, and the twist, and the thread guide
are reversed to the above. The ring frame seems to take
the lead for continuous spinning, on account of it requiring
less power than the cap or flyer frames, and the ends are
pieced up much handier on the ring frame, reducing skilled
labor to a minimum. The winding on of the thread is
done by the little traveller slipping, and letting a little of
of the twist go out at the same time, therefore, the twist
is not perfect, and is injurious to one of the best elements

in cotton spinning; the speed of spindle and bobbin are constant with the drawing, when they are properly connected.   In the throstle and cap frames, the twisting and drawing, and winding on are simultaneous with a constant speed, but the winding on is due to the bobbins slipping. The throstle frame excels all other frames by making a nicer, smooth, and rounder thread, through being held and wrapped with the flyer leg, preventing its fibres from being whizzed out, as it travels from the drawing rollers to bobbin, as is not the case with the other two frames, but the cap frames on this very account will show more elasticity of thread than either flyer or ring frame, and by its whirling the thread against the cap guides, send any loose moats or neps that are capable of being beat out with such force, that they fall out by this rapid beating on to the floor, at the same time it makes the fibres of cotton to stand out like mule yarn, and giving it a wooly appearance, these machines have always been run at high speeds, requiring great power to drive them, and will turn as many hanks per day off as the ring frame, but generally the yarn is not so well spun, even from the same roving as on the other frames, although it is yet admired by some for certain kinds of goods for warp yarn, on account of its elasticity acting favorably with the reed, when striking the weft up, and in shedding also, its spongy nature makes it a desirable class of yarn for the filling up of certain cloths, deriving a large percentage of gain over the wiry thread made on throstle frame, so you see the different methods of putting the twist in, makes different class of yarns, and all made from the same preparation, "hence the making of a warp by the mixture of these different yarns would be simply ridiculous, showing us that the quality of the yarn is not to be determined altogether by the treatment it receives before it reaches the spinning process, still this good treatment must be held up as a prevailing power towards acquiring an even thread. For these frames can claim no power of making that over the preparation that has already been given it, there may some injury be done here, if the drawing and twisting be not strictly attended to, and it is the superior manner of twisting in the throstle frames that makes claim over every other kind of continuous spinning yet extant, and the flyers on the roving frames claim the same advantage, when running at a moderate speed, by excluding the action of the air on the soft twisted rovings, if not prevented will raise the loose fibres, and make them stand out at every angle imaginable and simultaneously with twist are fastened in that position, having the appearance of a bottle brush, when such roving is made and passed along to the spinning frame, you can imagine the kinks and cuts how they are made by the rollers when such a roving is presented. This evil has been fully demonstrated by the Danforth list speeders, for in attenuating of such roving you will see as it leaves the front roller on the spinning frame these imperfections as the twist runs up, making thick and thin places, the twist running into this thread as the (square of the diameters inversely), therefore, making the evil still worse, and that is not all, it exhibits a want of knowledge of the business on those who have the management under their care, and I hope these remarks will be suggestive of using every effort to contribute for the making of a smooth roving, with all the fibres laid longitudinal, which will be conducive to making an even thread. Generally speaking *evenness* of a thread "depends on the amount of doublings subordinate to the equality of fibres and draughts, and the exacting of intervals." If

these are the essentials for making a perfect even thread,
then they form a rule, which we are to be governed by,
and if we are "to believe that all things are not unreasonable, and to hope all things not impossible," then I
think we can closely approximate perfection, in making
an even thread, for this rule is the exact formula given through
the whole procedure of this little work and by taking the
advice as is given progressively you will obtain the qualites
of a good even thread. We are pretty well satisfied on
the whole, of the different machines in use for making
warps, and the class of goods required in our market, that
the ring frame is the most profitable one to use, its production in quantity and the class of yarn it makes, seems
to answer our wants for which our kind of fabrics are
made from. It being a kind of go between the throstle
yarn and mule yarn, which suits us admirably either for
warp or filling, and which this frame can easily be adapted,
it being an invention of this country, and in its development has undergone a variety of improvements since the
orignal prototype was first presented to the public,
every device be it ever so trifling that has been
attached to the frame, has been held as a claim by the at-
tache, until some other novelty as superceded it, and so
these improvements keep on according to the progressive
ideas, and genius which our country produces, until we
have arrived now at such excellence in the construction of
them, that there seems to be but little difference now in
the quantity of work turned off, as to who the machine
maker may be, their choice being left to their own judgement, with regard to durability, and exactness in work-
manship, which all tends to determine the cost of the machine. Before leaving this frame, I must refer to the
draught again, being 7.4 the breakage for back and middle

will be the quotient of minus $\frac{1}{10}$ of the whole draught equal
1.12 multiplied by 6.66 equal 7.4, and the gearing up of these
draught wheels, should be placed all at the gearing end of
front roller, to prevent lost motion, and the weighting of
top·roller in front, will be about seven pounds, and the
same for back and middle rollers together, this will be
sufficient for No. 20's yarn, for single boss rollers, and
with respect to the number of traveller to be used, I could
not specify, for you have to regulate them according to
the material your yarn is composed of, and the construction of the frame, be sure to have the ring rail free and
steady, and the inside of guide wire over the centre of
spindle, also the ring rail lifters to be at perfect right-
angles with the rails, and preserving their equal distances,
the latter.

The mule is a spinning machine also, and is mostly
used for making filling, or weft, this yarn has a soft, and
downy appearance and feel, so well adapted for the filling
up the web of cloth, and for the making of soft hosiery
yarns, owing, this peculiarity to the manner by which the
twist is put in the yarn, there being a striking contrast,
with the same number of turns of twist betwixt this mule,
and ring yarns.   Yet I have heard of experts, that could
not distinguish one from the other.   I guarantee that
any blind man can make a distinction in a piece of cloth,
where half of warp is mule yarn, and the other ring yarn,
if he has the sense of feeling in his fingers, and would give
you convincing proof, by letting his finger stop within a
shot of it.   Now in taking the whole demand for yarns,
this class is far greater than any other, and can be produced cheaper, this, is a great advantage, and such a one
that tells where there are a great numbers of spindles at
work, especially when similar number of yarn can be made

from a lower grade of stock, having the propensity of stretching, by the gain of carriage over the drawing rollers, which assists in making a uniform thread, by accomplishing what the drawing rollers fails in, where there are short and soft spongy places caused by the fibres not being held to be attenuated, and when issuing forth from the rollers, their lofty and bulky form resists the portion of twist, and consequently gives the thinner part of the thread more than its portion; therefore, when the carriage is made to travel faster than the circumference of front roller, these bulky portions which are rendered soft by not receiving their ratio of twist, become reduced by the increased tension and getting their portion of twist from the thinner parts of the thread, which had taken up more than was allotted to them, and will, by the time it reaches to the full stretch become a pretty uniform thread, but will not compare to the uniformity, of twisting given by the throstle frame.

We must admit, that the mule can be adapted to a greater assortment of yarns than any other machine yet brought out, and will retain its supremacy, as a more desirable machine for those who are spinning for the world's market, for these class of yarns are more in demand than any other, made by continuous spinning, which fails in fully accomplishing the task of imitating mule yarn. It has not been my intention in writing up this little work to describe fully the motions of these different machines, and their mechanical appliances which are necessary to their construction, but to render such information that will assist those who have the task of adjusting them, and the material from which they are to manipulate into whatever they are calculated to produce, and helping to facilitate, by working on a system already laid down in the

schedule to which can be referred to, until you have become thoroughly acquainted with the rules from which
this schedule was originally prepared, and made in order
to save time and labor, where sudden changes become
necessary, and ought to be done scientifically, making you
master of your situation, and not a mere tyro, but a journey-
man, who theoretically and practically understands what
every motion requires, without attempting to experiment,
and this is the reason why the skilled artisan should be
recompensed for his knowledge of the business he follows,
over those makeshifts who are continually applying for
these situations, and are often engaged, as a favor, but at
a price consistent with their knowledge. This being an
aggravation to every one who has to depend on the quality
of his work, whether theirs will be remunerative. But, in
the near future it will become necessary for the benefit of
the employers, to ask from each overseer, a certified
diploma of his capabilities, before engaging him, showing
that he is no empiric, but a bona fide applicant, having
been examined by a professor or expert, who shall make
out his papers according to his capabilities for the situation
he has so earnestly solicited, and for every applicant the
professor must exact a fee corresponding with his position
and extent of his examination, and by a system of this
kind the employer will have no doubt of his theoretical
knowledge, which can be relied on by the signature of
the professor on his papers, he must also show some recommendation of his practical qualifications, that he is no
impostor.

I must resume the manner of calculating the twist and
draught on the mule, as I have shown on the other machines,
for example, I shall make use of the wheels usually sent
on Parr's mules, but these are from an old mule ; however,

they will answer our pupose so long as we can show the
method by which we get the result, it makes no difference
as to what kind of mule it is, they are all obtained in the
same manner. We want to get the draught wheel, or
change pinion, to produce 7.4 draught for No. 20's yarn,
made from 2.92 hank roving. The front roller wheel is
13 teeth, and the stud is 78 teeth, the back roller wheel 50
teeth, and the back and front rollers being one in. in di-
ameter so the $\frac{50}{7.4}$. $\frac{78}{13}$, equals 40 change pinion ; this is exclusive of gain of carriage, which should have about $1\frac{1}{2}$ in.
for No. 20's yarn, with an increase of $\frac{1}{4}$ in. for every five
Numbers. The number of turns of twist per inch for No. 20's,
called *weft twist*, equals 14.5 multiplied by 3.1416, equals
45.57 turns for front roller, one revolution.   On this front
roller is a 120 wheel, which runs into a 32 wheel and ——
required rim runs a 10 in. rim on the cylinder. The cylinder
is 6 in. in diameter and the warve is $\frac{7}{8}$ in. in diameter, so
$\frac{45.57}{1}, \frac{7}{48}, \frac{32}{120}, \frac{10}{1}$, equals 17.72 in. rim, number of turns, equals $\frac{120}{32}$,
$\frac{17.72}{10}, \frac{49}{7}$, equals 45.57 turns for one revolution of front roller,
when divided by 3.1416, which means the circumference
of unity or 3.1416)45.57(14.5 turns of spindle for one
inch of yarn delivered by the front roller. The back
shaft is driven by a 22 on front roller, into a 51 on men-
doza wheel.   On this is a 19, which drives a 58 on the
back shaft, equals $\frac{58}{19}, \frac{51}{22}$, equals 7 turns of front roller for
back shaft, one revolution ; but the shaft makes three
revolutions in one stretch and the length delivered by front
roller, equals 66 in., when the diameter of roller is 1 in.
because seven revolutions multiplied by 3.1416, equals
22 in., multiplied by 3 revolutions, equals 66 in., and
calling the speed of carriage four stretch per minute, equals
66 in. multiplied by 4, equals 264 in. per minute for intermittent spinning, and when multiplied by the number of

turns per inch, it equals 14.5 multiplied by 264 in. equals 3828 revolutions of spindle, this divided by the number of turns per spindle for one revolution of front roller, equals 45.57)3828(84 revolutions of front roller per minute, giving us a production in pound weight, by the rule, equals length in inches per minute multiplied by number of minutes worked per week, this divided by number of inches in one hank multiplied by number of yarn. Then in the first place we must deduct about ten per cent. for stoppages, etc., equals 60 minus 6 equals 54 working hours *nett*, equals 54 multiplied by 60 minutes, equals 3240 minutes, and the number of inches in 840 yards equals 840 multiplied by 36 equals 30240 in. in one hank, so we get $\frac{3240 \times 264}{30240 \times 20}$'s equals 1.41 pounds per week per spindle of No. 20's weft twist yarn. It is evident now that if we had put the same number of turns per inch in this yarn, our production would show 20 per cent. less than continuous spinning does, but this deficiency is made up when you take into consideration the investment, wages, finding, etc., and I think will maintain what I have previously stated, that we can spin for the market cheaper by the mule reckoning, on the kind of stock that can be used and the class of yarns made from it, will enter largely in making up the discrepancies, by reduced length with intermittent spinning.

No. 9 TABLE—MISCELLANEOUS.

Hanks	Interval	Grains per Yard	Weight on Roller	Draught	Hanks per 10 h.	Lbs. per Sp.	Rev. of Roller	Layers	Gauge Points	Diam'tr	Circum.
.052	1.5	160	60								
.062	1.44	134	50								
.125	1.40	67	40					3.5			
.25	1.36	33.3	30					5.			
.5	1.32	16.6	25					7.			
.75	1.28	11.	25					8.6			
1.	1.26	8.3	22	3.17	9.36	18.7	1¼ 185	10.	16.	1	3.1416
1.5	1.24	5.5	22	3.63	10.2	13.6	1⅛ 177	12.	14.2	1⅛	3.534
2.	1.22	4.16	20	4.	10.5	10.5	190	14.	12.8	1¼	3.927
2.5	1.21	3.3	20	4.5	11.5	7.66	194	15.7	11.6	1⅜	4.319
3.	1.20	2.77	18	5.04	10.7	5.35	185	17.	10.63	1.5	4.712
3.5	1.19	2.4	18	5.42	10.5	4.2	180	18.5	1.062	15.5	47.1
4.	1.18	2.08	18	5.76	9.6	3.2	165	19.6	1.03	16	48.7
4.5	1.17	1.85	16	6.07	9.3	2.65	158	21.	1.	17	50.
5.	1.16	1.68	16	6.34	9.	2.25	153	22.	.941	18	53.4
5.5	1.15	1.52	16	6.6	8.5	1.88	144	23.	.88	19	56.6
6.	1.14	1.39	16	6.84	8.5	1.7	143	24.	.842	20	59.7
8.	1.12	1.02	14	7.07	8.0	1.42	136	28.	.8	21	62.9
10.	1.10	.83	14	7.07	8.0	1.33	136	31.5	.765	22	66.
12	1.08	.69	14	7.26	8.0			34.	.73	23	69.
14.	1.06	.59	14					37.	.7	24	72.3
16.	1.04	.52	12					40.	.67		75.4
18.	1.02	.46	12					42.			
20.	1.	.42	10					44.			

Rev. of Roller brackets: 7 x 3½　8 x 4　10 x 5

We will now take a review of what has been said in this little work, respecting the mode of procedure by taking the schedule as a guide for the *speeds, weights,* measures, &c., which has been framed for the purpose of introducing into our cotton mills a system, which will show a more methodical manner of making our yarns, giving every manufacturer the same privilege of taking advantage of this course if he chooses, being one so unique in its mode of precedure and placing each individual manufacturer on the same footing as to how he should proceed to make any special number of yarn at the least cost, causing a universality in the prices, bringing competitors on equal terms, that they may meet each other more agreeable in the markets, and hasten a continued friendship which is more desirable than a wrangling disposition, endeavoring to crush each other by reduced prices, causing enmity and discord, but by adopting this schedule system it will alleviate this feeling in a measure calculated in the cost by wages, for these will have a constant tendency in making and producing them on equal grounds, and so with the buying of your cotton if you have money or credit. Then if your competitor takes advantage of you it will come from some other source, whereby he can reduce the general expenses of his mill, by a small attribute of acuteness, which is a very essential thing to have in the business. It is also necessary for your machinery to be able to produce the quantity of work per day as shown in the schedule for each machine, and that will be enough to keep you safe, and insure you a better quality, than an overproduction from increased speeds, for there is a limit to all the preparatory machines, which you cannot overcome with the machines now in present use. We often hear of great productions, but annual statistics will show them wanting, who has been trying to jump the fence, for this boundary is one of our safeguards, to prevent trespassing outside of the limits of the schedule ; our experience has proved beyond doubt that these quantities herein specified, warrants the best and most satisfactory results, when compared with the most exaggerated productions, which are frail and faulty, by the excess, possessing little value for the amount of labor and power consumed. It may be inferred by some of our readers, that the spreaders and cards are not sufficiently taxed, and might as well do more than the quantity specified, which they presume is a dead loss to the manufacturer, and would bring him to ruin. Now I hold the reverse opinion, and would not increase the capacity given, for they will do no more and do it right. I have not specified the amount of work for *underflat* card, which claims to do double the quantity of an ordinary flat card, my experience with them has not proved it, by no means used at present, and these are what has always produced good carding on every other kind of card in present use, I do believe this, if the underflats was dispensed with, and substitute workers and strippers in their place, would be a much superior manner of increasing the quantity, they would also get clear of the short fibres and rubbish which are held in by these flats, choking them all up to centre line, causing the card to make neps, by being surcharged, which cannot be avoided in their present position ; and I will venture to say that there would be more carding virtues in presenting the lap to the cylinder by the feed rollers on this card, than there is by the two rollers and the underflats combined, I mean the principle by which good carding can be better attained, in lieu of two rollers to carry the cotton, and the underflats to hold the residue which should have dropped out. It would be a decided improvement if the lap was so presented, for it is the most available place and position for the cylinder to execute such good work being better prepared for the flats to receive, for more of the extraneous matter would be driven out, for when you come to compare the slow motion of feed rollers, with the rapid revolving cylinder, in contrast, to the quicker surface speed of lickerin which has no hold on the filaments to allow them to be teased out like the feed roller has with the cylinder. So in its present state of construction I could not determine what its production would be of good carding. The *revolving* top flat card exhibits good principles of carding, and a greater quantity than ordinary flat card can do, but, requires great attention, and skill to keep them in order, and this article being scarce or hard to keep at the price, they prove unsatisfactory and such a piece of ingenuity has to be forsaken on account of a more industrious class of help to attend them. The *roller* card still maintains its reputation which it has held for a century, and has no signs of becoming a martyr to any of the recent innovations which has been introduced of late, its proportion of quality to quantity has no other competitor where they approximate so nearly, except the revolving flat card which comes up pretty closely and proved itself a pretty good match in the race. But its no matter what kind of card it is, or how it is constructed, if it executes good quality of carding, with simple adjustments to produce it, that we may require less skilled labor to perform this task, for our object is to reduce the cost, but not altogether, at the risk of injuring our machines by engaging such help as are not competent of doing ordinary card room services it will be much better to employ skilled persons and increase the production, by better care and attention on them. The railway head has received some improvements, and are acknowledged to be more complete, but not sufficient to warrant an even sliver. Yet we may look for some bright genius being struck with the idea before long, of having invented a device by which the rollers will be ready in advance, to correct any inequality of slivers issuing from the trough up to back roller of railway head, this would be a great achievement, deserving all the eulogy and emoluments, which a manufacturing people could bestow, for such an improvement on this useful piece of machinery. The drawing frame stands where it did thirty years ago, and has received no substantial improvement since the coiler was applied, with the exception of a few gim-cracks, whose novelties soon become a nuisance, and are taken off as bric-a-brac. I am glad to see that some of our machine makers, have of late, come to the conclusion, to place all the driving gears for the rollers at one end of drawing frame, a much wanted change necessary for the quality of the drawing sliver, which was deficient, caused by the lost motion and torsion, from a serpentine course of driving the rollers, from both ends of the frame, we are also deriving some benefit by the adoption of short frames, so that we can have our steel roller in one continuous length, discarding with jointing by short rollers. There is also a hazard of ruining your sliver, by using calenders with grooves in for condensing them, I have repeatedly been forced to replace them with plane surfaced rollers on account of the fibres getting cut up by a lateral movement of the rollers and the trumpet getting out of place, and should consider certainty of more importance than risk, believing that you will keep a vigilant watch over this important machine, from what I have previously stated about it, for you are sure to reap the benefit from the care and attention so earnestly desired, which, in a measure, gives you the profit and loss account here.

When we come to slubber, intermediate, and roving frames, we find an increase in production by having a more substantial spindle and bolster reducing the vibrations caused by the flyers getting out of balance, and the constant wear of spindle and bolster, by the incessant alternating motion of bobbin rail, which used to shake so violently when in its lowest position, making uneven roving at every revolution of flyer, by its eccentricity, it is seldom we see anything of that kind now, having improved the construction of the machines, which insures firmness, with a desire for durability, being elegant in design, and elaborate in finish, leaving nothing more to desire, except in that they should have proper care and attention, by those who are in charge of them, keeping them cleaned and oiled enabling them to turn off the required production, as well as giving them a delightful appearance, and in making an effort to put in practice these useful hints, you will discover, how much easier it is to accomplish the days work, by having everything in its place, and put in use at the proper time, showing that a masterly discipline over your help, puts confusion and irregularities at a discount, by the order and civility which your help have recognized the place to be an institute, where good conduct and decorum is taught, tending to promote their welfare according to their industry, instead of a workshop of gabbling drivellers, whose habits bring disorder, making waste by their slovenly manner of working, bringing the machinery to rack and ruin, consuming more power, with less production.   Such is the course of undisciplined help, and would be better to discharge those who are not capable of reform, in preference to making your room into a rag or junk shop, besides you are compelled to have the quantity of work off according to schedule in a proper manner, so

that the succeeding machines may be regularly supplied.
With these few words of advice to those who feel inclined
to adopt this system, hoping they will appreciate it for its
economy and simplicity.   I shall now leave you with my
best wishes.